# Serotonin in Major Psychiatric Disorders

# PROGRESS IN PSYCHIATRY
## Series

David Spiegel, M.D.
Series Editor

# Serotonin in Major Psychiatric Disorders

*Edited by*
*Emil F. Coccaro, M.D.*
*Dennis L. Murphy, M.D.*

American Psychiatric Press, Inc.

Washington, DC
London, England

*Note:* The authors have worked to ensure that all information in this book concerning drug dosages, schedules, and routes of administration is accurate as of the time of publication and consistent with standards set by the U.S. Food and Drug Administration and the general medical community. As medical research and practice advance, however, therapeutic standards may change. For this reason and because human and mechanical errors sometimes occur, we recommend that readers follow the advice of a physician who is directly involved in their care or the care of a member of their family.

Books published by the American Psychiatric Press, Inc., represent the views and opinions of the individual authors and do not necessarily represent the policies and opinions of the Press or the American Psychiatric Association.

Copyright © 1990 American Psychiatric Press, Inc.
ALL RIGHTS RESERVED
Manufactured in the United States of America
First Printing    93 92 91 90 4 3 2 1

American Psychiatric Press, Inc.
1400 K St., NW, Suite 1101
Washington, DC 20005

The paper used in this publication meets the minimum requirements of the American National Standard for Information Sciences—Permanence of Paper for Printed Library Materials, ANSI Z39.48–1984.                                    ∞

**Library of Congress Cataloging-in-Publication Data**

Serotonin in major psychiatric disorders / edited by Emil F. Coccaro, Dennis L. Murphy.
    p.   cm. — (Progress in psychiatry)
    Includes bibliographical references.
    ISBN 0-88048-292-3
    1. Mental illness — Pathophysiology. 2. Serotonin — Physiological effect. 3. Mental illness — Etiology.
    I. Coccaro, Emil F. II. Murphy, Dennis L. III. Series.
    [DNLM: 1. Mental Disorders — Physiopathology.
    2. Serotonin — physiology. QV 126 S48545]
    RC455.4.B5S47 1990
    616.89'071 — dc20
    DNLM/DLC
    for Library of Congress                                    89-18281
                                                 CIP

**British Library Cataloguing in Publication Data**

A CIP record is available from the British Library.

# Contents

# Contributors

**George M. Anderson, Ph.D.**
Director, Laboratory of Developmental Neurochemistry; Research Scientist, Yale Child Studies Center and Department of Laboratory Medicine, Yale University School of Medicine, New Haven, Connecticut

**Charanjit S. Aulakh, Ph.D.**
Staff Fellow, Section on Clinical Neuropharmacology, Laboratory of Clinical Science, National Institute of Mental Health, Bethesda, Maryland

**Harry A. Brandt, M.D.**
Chief, Unit on Eating Disorders, Intramural Research Program, National Institute of Mental Health, Bethesda, Maryland

**Timothy D. Brewerton, M.D.**
Director, Eating Disorders Program; Assistant Professor, Department of Psychiatry and Behavioral Sciences, Medical University of South Carolina, Charleston, South Carolina

**Dennis S. Charney, M.D.**
Chief, Psychiatry Service, West Haven VA Medical Center; Associate Professor, Department of Psychiatry, Yale University School of Medicine, New Haven, Connecticut

**Emil F. Coccaro, M.D.**
Director, Outpatient Department and Clinical Neuroscience Research Unit, Eastern Pennsylvania Psychiatric Institute; Associate Professor, Department of Psychiatry, Medical College of Pennsylvania, Philadelphia, Pennsylvania

**John G. Csernansky, M.D.**
Gregory Couch Professor of Psychiatry, Washington University School of Medicine, St. Louis, Missouri

**Kenneth L. Davis, M.D.**
Chairman, Department of Psychiatry; Professor, Departments of Psychiatry and Pharmacology, Mount Sinai School of Medicine, New York, New York

**Kym F. Faull, Ph.D.**
Director, Pasarow Analytical Neurochemistry Facility; Senior Research Associate, Department of Psychiatry and Behavioral Sciences, Stanford University School of Medicine, Palo Alto, California

**George R. Heninger, M.D.**
Director, Ribicoff Research Facilities, Connecticut Mental Health Center; Professor, Department of Psychiatry, Yale University School of Medicine, New Haven, Connecticut

**J. Dee Higley, Ph.D.**
Staff Fellow, Division of Intramural Clinical and Biological Research, National Institute on Alcohol Abuse and Alcoholism, Rockville, Maryland

**Thomas R. Insel, M.D.**
Staff Psychiatrist, Section on Comparative Studies of Brain and Behavior, Laboratory of Clinical Science, National Institute of Mental Health, Bethesda, Maryland

**David C. Jimerson, M.D.**
Director of Research, Department of Psychiatry, Beth Israel Hospital; Associate Professor, Department of Psychiatry, Harvard Medical School, Boston, Massachusetts

**John H. Krystal, M.D.**
Research Director, Schizophrenia Special Studies Unit, West Haven VA Medical Center; Assistant Professor, Department of Psychiatry, Yale University School of Medicine, New Haven, Connecticut

**Brian L. Lawlor, M.D.**
Director, Geropsychiatry Division; Assistant Professor, Department of Psychiatry, Mount Sinai School of Medicine, New York, New York

**Michael D. Lessem, M.D.**
Assistant Professor, Department of Psychiatry, University of Texas Medical School, Houston, Texas

**Markku Linnoila, M.D., Ph.D.**
Clinical Director, Division of Intramural Clinical and Biological Research, National Institute on Alcohol Abuse and Alcoholism, Rockville, Maryland

**J. John Mann, M.D.**
Director, Laboratory of Psychopharmacology, Western Psychiatric Institute and Clinic; Professor, Department of Psychiatry, University of Pittsburgh School of Medicine, Pittsburgh, Pennsylvania

**P. Anne McBride, M.D.**
Director, Division of Pediatric Mental Health, New York Hospital; Assistant Professor, Department of Psychiatry, Cornell University Medical College, New York, New York

**Alan M. Mellow, M.D., Ph.D.**
Director, Geropsychiatry Research, Ann Arbor VA Medical Center; Assistant Professor, Department of Psychiatry, University of Michigan Medical School, Ann Arbor, Michigan

**Dennis L. Murphy, M.D.**
Chief, Laboratory of Clinical Science, National Institute of Mental Health, Bethesda, Maryland

**Linda M. Nagy, M.D.**
Director, Anxiety Disorders Clinic, West Haven VA Medical Center, West Haven; Assistant Professor, Department of Psychiatry, Yale University School of Medicine, New Haven, Connecticut

**Kim R. Owen, M.D.**
Assistant Professor, Department of Psychiatry, Yale University School of Medicine, New Haven, Connecticut

**Michelle A. T. Pato, M.D.**
Guest Researcher, Laboratory of Clinical Science, National Institute of Mental Health, Bethesda, Maryland

**Margaret Poscher, M.D.**
Senior Resident, Department of Psychiatry, Stanford University School of Medicine, Palo Alto, California

**William Z. Potter, M.D., Ph.D.**
Chief, Section on Clinical Pharmacology, Laboratory of Clinical Science, National Institute of Mental Health, Bethesda, Maryland

**Alec Roy, M.B.**
Director, Affective Disorders Program, Hillside Hospital, Glen Oaks, New York

**Matthew V. Rudorfer, M.D.**
Staff Psychiatrist, Section on Clinical Pharmacology, Laboratory of Clinical Science, National Institute of Mental Health, Bethesda, Maryland

**Larry J. Siever, M.D.**
Director, Outpatient Division, Mount Sinai Medical Center; Associate Professor, Department of Psychiatry, Mount Sinai School of Medicine, New York, New York

**Trey Sunderland, M.D.**
Chief, Unit on Geriatric Psychopharmacology, Laboratory of Clinical Science, National Institute of Mental Health, Bethesda, Maryland

**Stephen J. Suomi, Ph.D.**
Chief, Laboratory of Comparative Ethology, National Institute of
Child Health and Human Development, Bethesda, Maryland

**Matti Virkkunen, M.D.**
Senior Lecturer, Division of Forensic Psychiatry, Department of
Psychiatry, University of Helsinki, Helsinki, Finland

**Scott W. Woods, M.D.**
Chief, Anxiety Disorder Clinic, Clinical Neuroscience Research Unit,
Ribicoff Research Facilities, Connecticut Mental Health Center;
Assistant Professor, Department of Psychiatry, Yale University
School of Medicine, New Haven, Connecticut

**Krystina M. Wozniak, M.D.**
Visiting Associate, National Institute on Alcohol Abuse and
Alcoholism, Rockville, Maryland

**Joseph Zohar, M.D.**
Staff Psychiatrist, Be'er Sheva' Mental Health Center, Israel

**Rachel C. Zohar-Kadouch, M.D.**
Staff Psychiatrist, Be'er Sheva' Mental Health Center, Israel

# Introduction to the Progress in Psychiatry Series

The *Progress in Psychiatry* Series is designed to capture in print the excitement that comes from assembling a diverse group of experts from various locations to examine in detail the newest information about a developing aspect of psychiatry. This series emerged as a collaboration between the American Psychiatric Association's (APA) Scientific Program Committee and the American Psychiatric Press, Inc. Great interest was generated by a number of the symposia presented each year at the APA Annual Meeting, and we realized that much of the information presented there, carefully assembled by people who are deeply immersed in a given area, would unfortunately not appear together in print. The symposia sessions at the Annual Meetings provide an unusual opportunity for experts who otherwise might not meet on the same platform to share their diverse viewpoints for a period of 3 hours. Some new themes are repeatedly reinforced and gain credence, while in other instances disagreements emerge, enabling the audience and now the reader to reach informed decisions about new directions in the field. The *Progress in Psychiatry* Series allows us to publish and capture some of the best of the symposia and thus provide an in-depth treatment of specific areas that might not otherwise be presented in broader review formats.

Psychiatry is by nature an interface discipline, combining the study of mind and brain, of individual and social environments, of the humane and the scientific. Therefore, progress in the field is rarely linear—it often comes from unexpected sources. Further, new developments emerge from an array of viewpoints that do not necessarily provide immediate agreement but rather expert examination of the issues. We intend to present innovative ideas and data that will enable you, the reader, to participate in this process.

We believe the *Progress in Psychiatry* Series will provide you with an opportunity to review timely new information in specific fields of interest as they are developing. We hope you find that the excitement of the presentations is captured in the written word and that this book proves to be informative and enjoyable reading.

David Spiegel, M.D.
Series Editor
*Progress in Psychiatry* Series

# Progress in Psychiatry Series Titles

# Introduction

The central serotonergic system has been hypothesized to play an important role in the regulation of a wide variety of neurobiological functions and behaviors. Over the past few decades evidence has accumulated to suggest that the central serotonergic system is involved in the regulation of appetite, sleep, pain, circadian rhythm, cognition, mood and anxiety states, and impulse control. While most serotonin (5-hydroxytryptamine [5-HT]) is concentrated in a cluster of midbrain nuclei, projections from these cells ramify, but variably, to all brain areas, including the frontal cortex, the striatum, and limbic system structures such as the hypothalamus and hippocampus. Thus, neuroanatomical evidence supports the notion that central 5-HT may play an important role in the regulation of a wide variety of the neurobiological systems that are integrated to serve as a substrate for mammalian behavior.

Increasingly, abnormalities of central serotonergic function are hypothesized to underlie behavioral disorders in which disturbances of these various neurobiological functions are observed. Thus, abnormalities of central 5-HT function are currently thought to play a significant role in disorders of mood, anxiety, cognition, eating, and impulse control, among others.

In 1987 an American Psychiatric Association Scientific Program symposium panel comprised of Drs. Coccaro (Chair), McBride (Co-Chair), Zohar, Brewerton, and Murphy presented data from a series of psychobiological studies of central serotonergic function so as to review evidence regarding the involvement of 5-HT in a variety of behavioral disorders. The success of the symposium led to the present opportunity to review similar data, in this *Progress in Psychiatry* series volume, concerning a more comprehensive range of disorders than could possibly be reviewed in symposium format.

Accordingly, this volume includes 10 chapters that review present strategies for the assessment of central serotonergic system function in humans (Chapter 1), as well as data regarding the various serotonergic hypotheses for developmental changes in behavior (Chapter 2), developmental disorders such as autism (Chapter 3), mood and personality disorders (Chapter 4), obsessive-compulsive disorder (Chapter 5), panic and generalized anxiety disorders (Chapter 6), eating disorders (Chapter 7), suicide, violence, and alcoholism (Chapter 8), schizophrenia (Chapter 9), and mechanisms of action of antidepressant and 5-HT-specific agents (Chapter 10). The final chapter (Chapter 11) reviews the clinical significance of central serotonergic system dysfunction in major psychiatric disorders.

This cross-sectional approach to the examination of central serotonergic system function in clinical psychobiology addresses the question of how a single neurotransmitter system may be abnormal in a variety of behavioral disorders and contribute to clinical phenomena that may appear to be uniquely different. The task of research in this field is to separate the similar threads running between these many disorders and attempt to determine how dysfunction in central serotonergic activities, which may seem apparently similar, can produce apparently different behavioral disorders. Such differences may be due to a variety of conditions, including differences in the extreme (i.e., hypofunction vs. hyperfunction), in receptor subtype-specific (e.g., $5\text{-HT}_{1A-C}$ vs. $5\text{-HT}_2$ or $5\text{-HT}_3$) function, in neuroanatomical subsystem function (e.g., cortical vs. hippocampal or hypothalamic), or in interactions with other neurotransmitter systems (e.g., noradrenergic, dopaminergic). Although data supporting any of these specific possibilities are limited at this time, it is clear that these hypotheses form the basis of where future studies can be expected to progress.

Emil F. Coccaro, M.D.
Dennis L. Murphy, M.D.

# Chapter 1

## *Strategies for the Study of Serotonin in Humans*

Dennis L. Murphy, M.D.
Alan M. Mellow, M.D., Ph.D.
Trey Sunderland, M.D.
Charanjit S. Aulakh, Ph.D.
Brian L. Lawlor, M.D.
Joseph Zohar, M.D.

# Chapter 1

# *Strategies for the Study of Serotonin in Humans*

T his chapter includes a brief review and critique of the approaches used to evaluate the status of serotonergic function in humans. The chapter is principally oriented toward the attempts to investigate possible associations between brain serotonin (5-hydroxytryptamine [5-HT]) and neuropsychiatric disorders, including changes in 5-HT–related measures during the treatment of these disorders. Some implications from new developments in 5-HT neurobiology and psychobiology that may contribute to the interpretation of present data and the design of future studies are also considered.

Serotonin pharmacology has undergone a remarkable renaissance in the last few years, witnessed, in part, by the publication of several recent monographs that provide valuable background information, primarily from animal studies, for developing new strategies for studying serotonergic function in humans (Green 1985; Rech and Gudelsky 1988). Some other monographs and reviews are valuable citation sources, particularly since the measurement of 5-HT and its metabolites in neuropsychiatric disorders and in relation to behavior extends back over 30 years (Barchas and Usdin 1973; Cowen 1987; Ho et al. 1982; Jacobs and Gelperin 1981; Meltzer and Lowy 1987; Murphy et al. 1978; Murphy et al. 1986).

## *DIRECT SEROTONIN MEASURES*

### Serotonin and 5-Hydroxyindoleacetic Acid

Concentrations of 5-HT in postmortem human brain samples, blood (where most 5-HT is contained in platelets), and urine have been measured in control subjects and patients with various neuropsychiatric disorders. Lumbar cerebrospinal fluid (CSF) contains very low mean 5-HT concentrations that, in all but individuals receiving monoamine oxidase inhibitors (MAOIs), are close to the detection limits (10 pg/ml) of current high-performance liquid chromatography (HPLC) assays (Anderson et al. 1990).

The principal 5-HT metabolite, 5-hydroxyindoleacetic acid (5-HIAA), has also been measured in human brain, CSF, plasma, and urine. Greatest interest has focused on CSF 5-HIAA as a possible index of central nervous system (CNS) serotonergic activity, with several hundred reports on CSF 5-HIAA in the literature.

### Serotonin in Neurons and Serotonin Binding Sites

Brain 5-HT neurons can be directly visualized by histofluorescence techniques. 5-HT cell bodies, which are almost exclusively located in the brain-stem raphe nuclei, have been quantitatively compared in human postmortem brain samples from control subject and neuropsychiatric patient groups (Mann et al. 1984; Yamamoto and Hirano 1985).

Similarly, some investigations of human brain 5-HT binding sites and putative receptors have been reported; $5\text{-}HT_1$ and $5\text{-}HT_2$ binding sites have been separately measured in control subject and patient groups using such ligands as $^3$H-labeled 5-HT, spiperone, and ketanserin (McKeith et al. 1987; Morgan et al. 1987). Platelets contain $5\text{-}HT_2$ (but not $5\text{-}HT_1$) binding sites, and there are an increasing number of reports investigating these receptors in neuropsychiatric patients (e.g., see Chapter 3).

As indicated in Table 1-1, a series of additional 5-HT binding sites have recently been identified and semiquantitatively localized by autoradiography in human brain. These sites have not yet been comparatively examined in patient groups. Of pertinence to some of the indirect studies of serotonergic function using 5-HT-selective agonists discussed in the subsequent portions of this chapter are the data showing that no $5\text{-}HT_{1B}$ sites are found in human brain (Heuring et al. 1986; Hoyer et al. 1986a). However, it is noteworthy that in the human brain areas corresponding to those areas in rodent brain where $5\text{-}HT_{1B}$ sites are densest, including the basal ganglia, there is a substantial number of incompletely identified $5\text{-}[^3H]HT_1$ binding sites that cannot be subcategorized on the basis of ligand colocalization comparisons as $5\text{-}HT_{1A}$, $5\text{-}HT_{1C}$, $5\text{-}HT_2$, or $5\text{-}HT_3$ sites. This suggests the possibility that there may be some human $5\text{-}HT_1$ site analogous, at least in neuroanatomical localization, to the rodent $5\text{-}HT_{1B}$ site. Recent but limited data have verified that more than 50% of the $5\text{-}[^3H]HT$ binding in the human caudate nucleus and substantia nigra represents a newly designated $5\text{-}HT_{1D}$ site (Peroutka et al. 1989; Waeber et al. 1988).

Positron-emission tomography has begun to be used to localize

**Table 1-1.** Localization of serotonin binding sites in human brain

| Binding sites | Cortex | Hippocampus | Amygdala | Caudate nucleus | Globus pallidus | Substantia nigra | Hypothalamus | Raphe nucleus | Cerebellum | Choroid plexus |
|---|---|---|---|---|---|---|---|---|---|---|
| 5-HT$_1$ | 350 | 400 | 275 | 225 | 550 | 725 | 275 | 600 | 15 | 450 |
| 5-HT$_{1A}$ | 300 | 350 | 200 | 10 | 15 | 3 | 70 | 400 | — | 20 |
| 5-HT$_{1B}$ | None | None | — | — | — | — | — | — | — | — |
| 5-HT$_{1C}$ | 25 | 100 | 50 | 100 | 140 | 200 | 100 | 50 | 15 | 1,700 |
| 5-HT$_{1D}$ | 81 | — | — | 190 | — | 329 | — | — | — | — |
| 5-HT$_2$ | 225 (EC) | 150 | 125 | 150 | 75 | 100 | 175 (MB) | 40 | 50 | — |

*Note.* Data are expressed as fmol/mg protein. Ligands and specific binding conditions were as follows: 5-HT$_1$, 5-[$^3$H]HT (1μM 5-HT); 5-HT$_{1A}$, $^3$H-labeled 8-hydroxy-2-(di-*n*-propylamino)tetralin (8-OH-DPAT) (1 μM 5-HT); 5-HT$_{1B}$, $^{125}$I-labeled cyanopindolol (30 μM isoproterenol); 5-HT$_{1C}$, $^3$H-labeled mesulergine (1 μM mianserin); 5-HT$_{1D}$, 5-[$^3$H]HT, 100 nM 8-OH-DPAT and 100 nM mesulergine (10 μM 5-HT); 5-HT$_2$, $^3$H-labeled ketanserin (1 μM mesulergine).

*Source.* Data are summarized from Hoyer et al. 1986a, 1986b; Pazos and Palacios 1985; Pazos et al. 1987; Waeber et al. 1988.

and begin to quantitate serotonergic components. $^{11}$C-labeled N-methylspiperone and N-methylbromolysergic acid diethylamide are being studied as 5-HT$_2$ receptor ligands in healthy subjects (Wong et al. 1984, 1987). Studies of patient groups have not yet been reported.

## Technical Considerations Regarding Some Direct Serotonin Assessment Measures

Determinations of 5-HIAA and 5-HT in human brain and in most bodily fluids have now been very reliably accomplished. Many separate metabolic steps, however, are involved in the formation, storage, and release of 5-HT and 5-HIAA. Dietary changes in intake of carbohydrates, of 5-HT precursor L-tryptophan, or of neutral amino acids  that compete with L-tryptophan for transport into brain, can change brain 5-HT and 5-HIAA concentrations. In addition, there is considerable evidence for functionally and nonfunctionally related pools of 5-HT (and, correspondingly, of 5-HIAA), exemplified in a recent study measuring alterations in the firing rates of 5-HT–containing dorsal raphe neurons (Trulson 1985).

As expected from animal studies, 5-HIAA measured in human CSF has been shown to be sensitive to drugs affecting 5-HT synthesis (L-tryptophan, p-chlorophenylalanine), release (fenfluramine), uptake (tricyclics), and metabolism (MAOIs). Such changes are very helpful in assessing drug effects, but at the same time provide evidence that 5-HT or 5-HIAA differences found in some neuropsychiatric disorders necessarily reflect the summed effects of many processes. It is possible, for instance, that altered CSF 5-HIAA might reflect a gross abnormality in 5-HT synthesis and/or storage, or a change in the total number of serotonergic neurons. However, functional consequences cannot readily be inferred from a single set of CSF 5-HIAA values, and other studies are needed to identify what processes are involved in any 5-HIAA difference.

Platelet studies provide an opportunity to examine some dynamic aspects of serotonergic function, such as 5-HT transport, in living human cells. A genetically based malfunction in 5-HT uptake, storage, release, or receptor characteristics might be found both in platelets, which are neuroectodermal derivatives, and in brain. Alternatively, these same platelet functions may be affected in a disorder by humoral factors acting on both brain and platelets; thus, platelets may provide an easily obtainable tissue for studies relevant to brain function.

A disadvantage of the platelet model is that these cells do not have neuronal cell-to-cell connections and hence are not subject to the same regulatory influences as are 5-HT neurons in brain. They may, in fact, be subject to plasma constituents (e.g., increased epinephrine in anxiety or stress situations) that are known to affect platelet aggregation and life span. Because 5-HT content, 5-HT uptake, and many other 5-HT–related functions in platelets differ in relation to platelet density and size, which are functions of platelet life span (Corash et al. 1984), apparent changes in these 5-HT–related functions may occur independently of changes in the brain; thus, there are some limitations to conclusions that can be drawn from platelet studies alone. Of considerable interest, some correlations between platelet 5-HT2 receptor numbers and pharmacologic responsivity to 5-HT have been observed (see Chapter 3). Similarly, there is some evidence suggesting associations between CSF 5-HIAA and neuroendocrine responsivity to a 5-HT agonist (Koyama et al. 1987).

## INDIRECT ASSESSMENT OF THE FUNCTIONAL STATUS OF HUMAN BRAIN SEROTONERGIC SYSTEMS

Several different indirect approaches are being employed to evaluate some functional components of the serotonergic systems in humans. Derived, in part, from the precursor load strategies that are used to assess the status of metabolic pathways, various responses to the 5-HT precursors L-tryptophan and 5-hydroxytryptophan (5-HTP) have been measured. Drugs with selective serotonergic actions have also been studied. Most commonly, neuroendocrine measures (e.g., changes in plasma prolactin, cortisol, or ACTH concentrations) have been used as end points (Meltzer et al. 1982a, 1982b; Murphy et al. 1986; Tuomisto and Mannisto 1985). Because these neuroendocrine responses appear to be mediated by brain serotonergic mechanisms, it has been proposed that the study of the neuroendocrine consequences (namely, plasma hormone changes) of manipulations of the serotonergic system (namely, peripheral administration of centrally active serotonergic agents) may reflect central 5-HT function and may be used directly as a tool for elucidating the neurochemical underpinnings of certain psychiatric disorders in humans. In this way, neuroendocrine challenge strategies have been viewed as "probes" of central 5-HT function; the validity of this approach, of course, depends upon the specificity of the challenge agents used and the responses measured (Meltzer et al. 1982a, 1982b; Murphy et al. 1986, in press; van Praag et al. 1987).

### Prolactin Responses

In rodents, nonhuman primates, and humans, plasma prolactin increases following serotonergic stimulation. Normally, prolactin release from the pituitary is tonically controlled by an inhibitory action of dopamine (Leong et al. 1983). Sustained elevations of plasma prolactin have been demonstrated after direct intraventricular injection of 5-HT (Krulich et al. 1979), and after administration of 5-HT precursors (Lamberts and MacLeod 1978; Meltzer et al. 1982b) as well as direct-acting 5-HT agonists (Fuller et al. 1981). The ability of 5-HT to release prolactin appears not to involve a direct effect on pituitary lactotrophs (Lamberts and MacLeod 1978). There is some evidence that 5-HT acts to actively release prolactin, rather than to disrupt tonic dopaminergic inhibition, although this point remains controversial (Clemens et al. 1978). Serotonergic projections arising from the midbrain raphe nuclei projecting into the mediobasal hypothalamus have been proposed as the anatomical pathway mediating this release of prolactin (Preziosi 1983).

## Cortisol and ACTH Responses

The role of 5-HT in the regulation of pituitary-adrenocortical activity is well established; however, the situation may be more complex than is the case with prolactin release, and many details remain to be clarified. Several studies in rodents have demonstrated fluctuations in brain 5-HT levels that correspond to the diurnal rhythm of corticosteroid secretion; this rhythm can be abolished by central 5-HT depletion or neurotoxic destruction of serotonergic pathways (Scapagnini et al. 1974).

With regard to acute effects of serotonergic manipulation, most studies support a stimulatory effect on the hypothalamo-pituitary-adrenocortical axis (Fuller 1981). 5-HT can promote the release of corticotropin-releasing factor (CRF) from rat hypothalamus in vitro (Buckingham and Hodges 1979). Studies using peripheral administration of serotonergic agents have, for the most part, demonstrated elevation of plasma corticosteroid levels. This appears to hold for 5-HT precursors and uptake inhibitors, as well as direct-acting 5-HT receptor agonists (Fuller 1981). In a recent study in which the effects of several selective 5-HT agonists and antagonists were examined, it was suggested that serotonergic corticosteroid release may be mediated through a 5-HT$_{1A}$ receptor (Lorens and Van de Kar 1987). Regarding the anatomical site of serotonergic action in stimulating adrenocortical function, an effect on hypothalamic neurons separate from those mediating prolactin release has been suggested (Van de Kar et al. 1985). Noradrenergic contributions and a nonhypothalamic component for this action of serotonergic agents have also been proposed (Smythe et al. 1988).

## Neuroendocrine Responses to Serotonin Precursors in Healthy Human Subjects

*L-Tryptophan.* While orally administered tryptophan has variable effects on plasma prolactin, intravenously administered L-tryptophan (5–10 g, or up to 100 mg/kg) caused significant elevations in plasma prolactin over baseline in three studies (Charney et al. 1982; MacIndoe and Turkington 1973; Winokur et al. 1986). Pretreatment with the 5-HT antagonists methysergide or metergoline reduced the effect of L-tryptophan on plasma prolactin secretion (MacIndoe and Turkington 1973; McCance et al. 1987). Furthermore, acute pretreatment with the tricyclic antidepressant clomipramine (a relatively selective 5-HT reuptake blocker) enhanced the prolactin response to L-tryptophan (Anderson and Cowen 1986). These findings support the hypothesis that in humans the plasma prolactin rise after administration of tryptophan is mediated by a serotonergic mechanism. In one study (Cowen and Anderson 1986), the $5\text{-}HT_2$ antagonist ketanserin failed to inhibit the prolactin response to L-tryptophan, suggesting that in humans this effect is mediated by a $5\text{-}HT_1$, rather than a $5\text{-}HT_2$, receptor subtype. In another study (Charig et al. 1986), ritanserin, a different selective $5\text{-}HT_2$ antagonist, enhanced the prolactin response to L-tryptophan, suggesting the possibility of a modulatory interaction between the $5\text{-}HT_1$ and $5\text{-}HT_2$ receptor subtypes. This effect was not sustained when ritanserin was given for 2 weeks (Idzikowski et al. 1987). An attenuation of the prolactin response to L-tryptophan by acutely, but not chronically, administered diazepam was interpreted as being suggestive of serotonergic involvement in the immediate therapeutic effects of this antianxiety agent (Nutt and Cowen 1987).

Changes in levels of cortisol and ACTH have also been evaluated after the administration of L-tryptophan, but highly variable results have been obtained in different studies. The diurnal variation of these hormones makes comparison of the effects of L-tryptophan with placebo imperative. Because several other serotonergic agonists increase levels of cortisol and ACTH, further studies are warranted.

Increases in growth hormone after L-tryptophan administration have been shown in some studies (Charney et al. 1982; Koulu 1982; Winokur et al. 1986). Growth hormone changes were inhibited by pretreatment with the nonselective 5-HT antagonist cyproheptadine (Fraser et al. 1979) and potentiated by pretreatment with the 5-HT reuptake blocker clomipramine (Anderson and Cowen 1986). Both growth hormone and prolactin responses were increased following food intake restriction (a 1,200 calorie diet for 3 weeks) in healthy volunteers, indicating the importance of controlling for diet in patient–control subject comparisons (Goodwin et al. 1987).

*5-Hydroxytryptophan.* L-Tryptophan can be converted to 5-HTP only by the tryptophan hydroxylase present in serotonergic neurons; thus, L-tryptophan has fairly selective actions. 5-HTP, the immediate precursor of 5-HT, can be decarboxylated by the nonspecific aromatic L-amino acid decarboxylase that is present in many monoamine systems and in other cells; hence, the specificity of 5-HTP as a 5-HT agonist challenge agent has been questioned. Nonetheless, 5-HTP administered intravenously to subjects pretreated with a decar-boxylase inhibitor has been shown to cause a significant increase in prolactin levels (Lancranjan et al. 1977). After being orally administered, 5-HTP has been variably reported to increase plasma prolactin levels (Westenberg et al. 1982).

As with L-tryptophan, the effects of 5-HTP on cortisol and ACTH have not been well studied, and existing results are contradictory. In one investigation (Imura et al. 1973), a significant increase above baseline in levels of both ACTH and cortisol occurred after 5-HTP administration in almost all subjects tested. In contrast, Westenberg et al. (1982) were unable to confirm a significant change in cortisol, and Meltzer et al. (1984a) found significant increases only in patients but not in control subjects. However, interesting associations between cortisol responses to 5-HTP and clinical features, as well as changes in responses during treatment with lithium and MAOIs, suggest that further studies would be of interest (Meltzer et al. 1984a, 1984b, 1984c). Apparently, studies with 5-HT antagonists have not yet been accomplished.

In regard to growth hormone, studies of the effects of orally administered 5-HTP have yielded mixed results: some groups have reported a significant increase in growth hormone over baseline, but others have not (Imura et al. 1973; Nakai et al. 1974). The only study including a full placebo day revealed no significant difference between 5-HTP and placebo (Handwerger et al. 1975). The effects of intravenously administered 5-HTP on growth hormone have also been studied. In subjects pretreated for 3 days with a decarboxylase inhibitor, intravenous infusions of 5-HTP led to a significant increase in growth hormone over baseline, an increase that was also significantly greater than that following a placebo infusion (Lancranjan et al. 1977). Cyproheptadine has been reported to diminish the increase in growth hormone following 5-HTP administration (Nakai et al. 1974).

## Neuroendocrine Responses to Selective Serotonin Agonists in Healthy Humans

*Fenfluramine.* Fenfluramine is an anorectic agent similar in structure to amphetamine but without stimulant properties. Biochemical data indicate that the 5-HT agonist properties of fenfluramine primarily derive from its ability to promote a rapid release of 5-HT (Kannengiesser et al. 1976). Fenfluramine can also inhibit 5-HT reuptake, and receptor-binding studies suggest that both fenfluramine and its principal metabolite, norfenfluramine, may have direct, albeit weak, 5-HT receptor agonist activity (Garattini et al. 1975, 1979).

Neuroendocrine studies have shown that fenfluramine administration leads to increased serum prolactin levels in healthy humans (Casanueva et al. 1984; Quattrone et al. 1983; Siever et al. 1984). The elevations in prolactin levels were proportionate to dose within the range of 20–100 mg of fenfluramine administered orally; statistically significant elevations occurred in subjects administered 60, 80, or 100 mg (Quattrone et al. 1983). Metergoline, a $5\text{-HT}_1/5\text{-HT}_2$ receptor antagonist, completely blocked the prolactin increase produced by 60 mg of fenfluramine (Quattrone et al. 1983).

Fenfluramine also increases levels of plasma cortisol and ACTH (Lewis and Sherman 1984). In one study, the increases in levels of cortisol and ACTH were proportionate to fenfluramine doses within the range of 0.5–1.5 mg/kg and were significantly blunted by pretreatment with cyproheptadine (Lewis and Sherman 1984). Fenfluramine also has been shown to augment the cortisol response to insulin-induced hypoglycemia; however, fenfluramine administered intravenously did not significantly alter growth hormone concentrations in the same subjects who demonstrated marked elevations in levels of prolactin (Casanueva et al. 1984).

*m-Chlorophenylpiperazine.* An aryl-substituted piperazine that is a direct-acting 5-HT receptor agonist, *m*-chlorophenylpiperazine (m-CPP) (a metabolite of the antidepressant trazodone), has recently been shown to produce dose-dependent elevations in plasma levels of prolactin in healthy control subjects (Mueller et al. 1985b). Although early studies in rodents suggested a preferential action of m-CPP on $5\text{-HT}_{1B}$ sites (Sills et al. 1984), m-CPP binds to multiple 5-HT and other brain neurotransmitter sites (Hamik and Peroutka 1989; Hoyer 1988), and may have a more complex mode of action (Murphy et al. 1989). m-CPP administered orally led to highly statistically significant elevations in plasma levels of prolactin and cortisol and to significant but small increases in body temperature and in several mood ratings, including euphoria and anxiety in healthy subjects (Mueller et al. 1985a). Similar responses to m-CPP were also observed after intravenous administration of this agent (Charney et al. 1987; Murphy et al. 1989). The changes in levels of prolactin and cortisol, as well as the temperature changes, induced by m-CPP in humans and in rhesus monkeys were blocked by pretreatment with metergoline (Aloi et al. 1984; Mueller et al. 1986).

Quipazine and MK-212, two structurally related piperzines, have been less studied. In one study in healthy humans, quipazine led to elevations in levels of cortisol but not of prolactin, growth hormone, or other pituitary hormones; gastrointestinal side effects also occurred (Parati et al. 1980). MK-212 increased plasma levels of prolactin and cortisol but did not change growth hormone concentrations (Lowy and Meltzer 1988).

In rodents, long-term treatment with antidepressants such as tricyclics, MAOIs, and lithium has been reported to alter m-CPP-induced neuroendocrine responses, temperature, food intake, and other behavioral responses as well as produce changes in 5-HT$_1$ and 5-HT$_2$ receptor densities (Anderson 1983; Aulakh et al. 1987, 1988, 1989; Cohen et al. 1983; Treiser et al. 1981; Zohar et al. 1988).

*Other Agents.* The serotonergic antianxiety agent buspirone, which selectively binds to 5-HT$_{1A}$ sites in rodents (Goa and Ward 1986), administered orally led to elevations in plasma levels of prolactin and growth hormone in healthy subjects (Meltzer et al. 1983). The indole hallucinogen $N,N$-dimethyltryptamine also elevated plasma prolactin and growth hormone in drug-experienced volunteers (Meltzer et al. 1982a). The serotonin selective reuptake inhibitor clomipramine, administered intravenously, increased plasma prolactin (Laakman et al. 1984); one other 5-HT reuptake inhibitor increased plasma cortisol and beta-endorphin (Petraglia et al. 1984).

## Other Physiologic Responses to Serotonin Agonists

While neuroendocrine changes have comprised the most widely used end points in pharmacologic challenge evaluations of the functional status of brain 5-HT in humans, some other measures are beginning to be more widely employed.

*Cardiovascular Changes.* Small increases in blood pressure and heart rate follow administration of some serotonergic agonists, including intravenously administered L-tryptophan and m-CPP, but these increases have not been prominent with other agents, including

5-HTP and fenfluramine. Cardiovascular changes most likely represent a combined consequence of peripheral and central 5-HT–mediated events. For example, in pithed rats, a predominant peripheral vascular 5-HT$_2$–mediated vasoconstriction appeared to be the dominant factor in the dose-related hypertensive actions of m-CPP (Bagdy et al. 1987). Of interest, the model 5-HT$_{1A}$ agonist 8-hydroxy-2-(di-$n$-propylamino)tetralin (8-OH-DPAT) has hypotensive properties in rodents (Fozard et al. 1987).

***Temperature Responses.*** Temperature is a multineurotransmitter-regulated function that can be modulated by serotonergic drugs. In rodents central serotonergic systems are known to be associated with thermoregulation, and most of the evidence points to the preoptic area of the anterior hypothalamus as playing the most important role (Hardy 1961). However, the exact nature of the response to 5-HT is controversial (Cox and Lomax 1977). Both hypothermia (Cox and Lee 1981) and hyperthermia (Myers 1981) have been induced in rats by 5-HT or 5-HT agonists. It has been proposed that the differential effects of 5-HT or 5-HT agonists on rat body temperature depend on the exact agent and route of administration employed. The receptor subtypes involved in thermoregulation have only been partially clarified, although recent theories have suggested that activation of 5-HT$_{1A}$ receptors by 5-HT agonists may lead to decreases in body temperature, whereas increases are mediated via activation of 5-HT$_{1B}$ and 5-HT$_2$ receptors (Gudelsky et al. 1986). Such opposing temperature responses following activation of different 5-HT receptor subtypes may also account for the earlier discordant findings, with different 5-HT agonists acting predominantly on one or another receptor subtype.

In humans, reliable increases in body temperature follow orally and intravenously administered m-CPP (Mueller et al. 1985a; Murphy et al. 1989). The temperature increases produced by oral m-CPP are blocked by low doses of the 5-HT$_1$/5-HT$_2$ antagonist metergoline (Mueller et al. 1986). These increases are also attenuated during long-term treatment with the 5-HT–selective reuptake inhibitor clomipramine, in keeping with the hypothesis derived from animal studies that such long-term treatment produces a down-regulation of 5-HT receptors (Zohar et al. 1988).

***Behavioral Responses.*** Behavioral changes were not originally considered to be an end point in the assessment of brain 5-HT function by pharmacologic challenges in humans because only negligible or very subtle self-rated feeling states or observer-related behaviors were noted in the original studies with L-tryptophan, with 5-HTP alone or in combination with a decarboxylase inhibitor, or with fenfluramine. Initial studies with m-CPP administered orally also showed only slight and individually variable changes in anxiety and euphoria (Mueller et al. 1985a). However, when considerably more prominent anxiogenic and other behavioral responses were observed after intravenously administered m-CPP (Charney et al. 1987; Murphy et al. 1989), and when patient groups also appeared behaviorally supersensitive to orally administered m-CPP (Zohar et al. 1987), considerably greater interest developed in the behavioral consequences of selective 5-HT agonists and antagonists. The newer 5-HT1A agonists, including buspirone, gepirone, and ipsapirone, have antianxiety effects in humans during long-term administration (Goa and Ward 1986). Thus, as with temperature and blood pressure, there are behavioral data to suggest some contrary actions of the different 5-HT1–selective agonists.

Many serotonergic agonists given in larger doses to rodents have long been known to produce behavioral changes ranging from the variants of the so-called 5-HT syndrome, which is also accompanied by hyperactivity, to hypoactivity and altered performance on complex behavioral tasks (for reviews, see Green 1984; Jacobs and Gelperin 1981). (Behavioral changes in animals and in some neuropsychiatric patient populations in response to serotonergic agents are described in many of the chapters of this volume.)

## AN OVERVIEW AND SOME CONCLUSIONS

Several additional points and examples from studies in humans and animals deserve consideration in evaluating current and possible future strategies for studying 5-HT function in humans.

Increasing evidence indicates that the brain serotonergic system in rodents and, on the basis of more fragmentary data, in humans is heterogenous. The 5-HT–containing cell bodies in the raphe nuclei in the brain stem not only send efferent axons to different brain areas, but also have different modulatory afferents (Molliver 1987). Each nucleus appears capable of functioning autonomously or at least semi-independently from the others, and each is capable of responding differently to the same groups of drugs (e.g., 8-OH-DPAT or ipsapirone vs. trifluoromethylpiperazine, a congener of m-CPP) (Sinton and Fallon 1988). At 5-HT terminals, multiple 5-HT binding sites or receptors have been identified in human and rodent brain. Studies with selective agonists and antagonists indicate that different sites that are responsive to 5-HT are quite capable of having opposing actions. In addition, inhibitory and excitatory serotonergic pathways have been identified electrophysiologically.

It thus no longer seems possible to consider that a coherently acting "serotonergic system" exists in brain, but rather that a group of serotonergic systems is involved. An example of how many different (but also, in some cases, similar) "serotonergic" responses have recently been defined comes from comparisons of the effects of some $5\text{-}HT_{1A}$ vs. $5\text{-}HT_{1B}$ selective agents, as summarized in Table 1-2. There are similar data indicating that 5-HT agonists acting at $5\text{-}HT_2$ sites have opposite effects from those of 5-HT agonists acting at $5\text{-}HT_1$ sites. In addition, $5\text{-}HT_3$ sites appear to subserve an independently acting system that is also capable of modulating psychomotor activity, feeding, and other behaviors independently of $5\text{-}HT_1$ or $5\text{-}HT_2$ sites. A direct implication of these new data for studies in humans is that it no longer seems possible to consider single measures such as that of 5-HT or 5-HIAA content as valid indices of 5-HT activity relevant to any specific functional change, because without knowing which raphe nucleus subsystem is changed, and which 5-HT receptor site is being affected, reasonable conclusions cannot be drawn.

The data in Table 1-2 also serve to emphasize another point of concern to a major thrust of current serotonergic studies in humans regarding neuroendocrine end points. The information reviewed above and much other data indicate that neuroendocrine responses to 5-HT agonists or to drugs such as 5-HT–selective antidepressants that modulate serotonergic neurons may, but need not, parallel behavioral, temperature, cardiovascular, or other physiologic systems modulated by serotonergic neurons (e.g., Zohar et al. 1988). Thus, neuroendocrine response measures alone may not suffice as end points from which conclusions about 5-HT-mediated behavioral changes can validly be made.

Many nonspecific factors are capable of influencing measurements of 5-HT, 5-HIAA, 5-HT receptors, and responses to 5-HT precursors as well as selective 5-HT agonists and antagonists. Sex-related differences in serotonergically mediated processes have frequently been found, with females generally being more responsive. Many examples of diet-related alterations have been observed. Reductions with age have been reported both in 5-HT receptors measured in postmortem brain samples (Morgan et al. 1987) and by brain imaging techniques (Wong et al. 1984), and also in responses to 5-HT agonists (Lawlor et al. 1989).

An additional issue is that of species differences. Not only is a rodent-like $5\text{-}HT_{1B}$ site not found in human brain (or in bovine, guinea pig, and some other species), but, as might well be expected, many other species differences in 5-HT neuroanatomy, physiology, and pharmacology are known. One pharmacologic example pertinent to studies described in other chapters concerns responses to pharmacologic challenges in rodents vs. humans treated chronically with clomipramine in comparable doses. Patients with obsessive-compulsive disorder were found to manifest diminished or unchanged prolac-

**Table 1-2.**  Some similarities and differences between serotonin agonists of the $5\text{-}HT_{1A}$ and $5\text{-}HT_{1B}$ types

|  | $5\text{-}HT_{1A}$ (8-OH-DPAT, buspirone, gepirone) | $5\text{-}HT_{1B}$ (m-CPP, TFMPP) |
|---|---|---|
| High affinity for $5\text{-}[^3H]HT$ binding sites in rodent brain | Yes | Yes |
| Increases plasma prolactin and/or cortisol in humans and rodents | Yes | Yes |
| Decreases raphe firing rates | Yes | Yes |
| —via raphe autoreceptors | Yes | No |
| —via presynaptic terminal autoreceptors | No | Yes |
| Reduces 5-HT synthesis and 5-HIAA concentrations | Yes | Yes |
| Feeding in food-deprived rodents | Decrease | Decrease |
| Feeding in nonfood-deprived rodents | Increase | Decrease |
| "Serotonin syndrome" of motor behaviors and hyperactivity in rodents | Yes/no | No |
| Hypoactivity, sedation in rodents and monkeys | No | Yes |
| Temperature | Hypothermia | Hyperthermia |
| Penile erections in rodents and monkeys | No | Yes |
| Blood pressure in rodents | Decrease | Increase |
| Anxiety in humans | Anxiolytic | Anxiogenic |

Note.  8-OH-DPAT = 8-hydroxy-2-(di-$n$-propylamino)tetralin; m-CPP = $m$-chlorophenylpiperazine; TFMPP = trifluoromethylphenylpiperazine

tin responses and unchanged cortisol responses to m-CPP in paired comparisons with values obtained prior to clomipramine administration. In contrast, rats developed exaggerated prolactin responses but attenuated cortisol responses during chronic clomipramine administration (Wozniak et al. 1988; Zohar et al. 1988). While it might be extremely interesting if these differences should prove to be disorder-related rather than species-dependent, many other examples in the literature continue to indicate that extrapolations across species require great caution, and that studies in humans are needed to understand serotonergic physiology and pharmacology in healthy subjects and, by definition, in neuropsychiatric patients.

# REFERENCES

Aloi J, Insel T, Mueller E, et al: Neuroendocrine and behavioral effects of *m*-chlorophenylpiperazine administration in rhesus monkeys. Life Sci 34:1325–1331, 1984

Anderson GM, Mefford IN, Tolliver TJ, et al: Serotonin in human lumbar cerebrospinal fluid: a reassessment. Life Sci 46:247–255, 1990

Anderson IM, Cowen PJ: Clomipramine enhances prolactin and growth hormone responses to L-tryptophan. Psychopharmacology (Berlin) 89:131–133, 1986

Anderson JL: Serotonergic receptor changes after chronic antidepressant treatments: ligand binding, electrophysiological and behavioral studies. Life Sci 32:1791–1801, 1983

Aulakh CS, Cohen RM, Hill JL, et al: Long-term imipramine treatment enhances locomotor and food intake suppressant effects of *m*-chlorophenylpiperazine in rats. Br J Pharmacol 91:747–752, 1987

Aulakh CS, Zohar J, Wozniak KM, et al: Clorgyline treatment differentially effects *m*-chlorophenylpiperazine-induced neuroendocrine changes. Eur J Pharmacol 150:239–246, 1988

Aulakh CS, Haass M, Zohar J, et al: Long-term imipramine treatment potentiates m-CPP-induced changes in prolactin but not corticosterone and growth hormone in rats. Pharmacol Biochem Behav 32:37–42, 1989

Bagdy G, Szemeredi K, Zukowska-Grojec Z, et al: *m*-Chlorophenyl-piperazine increases blood pressure and heart rate in pithed and conscious rats. Life Sci 41:775–782, 1987

Barchas J, Usdin E (eds): Serotonin and Behavior. New York, Academic Press, 1973

Buckingham JC, Hodges JR: Hypothalamic receptors influencing the secretion of corticotropin releasing factor in the rat. J Physiol 290:421–431, 1979

Casanueva FF, Villanueva L, Penalva A, et al: Depending on the stimulus, central serotonergic activation by fenfluramine blocks or does not alter growth hormone secretion in man. Neuroendocrinology 38:302–308, 1984

Charig EM, Anderson IM, Robinson JM, et al: L-Tryptophan and prolactin release: evidence for interaction between 5-HT$_1$ and 5-HT$_2$ receptors. Hum Psychopharmacol 1:93–97, 1986

Charney DS, Heninger GR: Serotonin function in panic disorders: the effect of intravenous tryptophan in healthy subjects and panic disorder patients before and during alprazolam treatment. Arch Gen Psychiatry 43:1059–1065, 1986

Charney DS, Heninger GR, Reinhard JF, et al: The effect of intravenous L-tryptophan on prolactin and growth hormone and mood in healthy subjects. Psychopharmacology (Berlin) 77:217–220, 1982

Charney DS, Woods SW, Goodman WK, et al: Serotonin function in anxiety. II. Effects of the serotonin agonist m-CPP in panic disorder patients and healthy subjects. Psychopharmacology (Berlin) 92:14–24, 1987

Clemens JA, Roush ME, Fuller RW: Evidence that serotonin neurons stimulate secretion of prolactin releasing factor. Life Sci 22:2209–2214, 1978

Cohen RM, Aulakh CS, Murphy DL: Long-term clorgyline treatment antagonizes the eating and motor function responses to m-chlorophenylpiperazine. Eur J Pharmacol 94:175–179, 1983

Corash L, Costa JL, Shafer B, et al: Heterogeneity of human whole blood platelet subpopulations. III. Density-dependent differences in subcellular constituents. Blood 64:185–193, 1984

Cowen PJ: Psychotropic drugs and human 5-HT neuroendocrinology. Trends in Pharmacological Sciences 8:105–108, 1987

Cowen PJ, Anderson IM: 5-HT neuroendocrinology, in Recent Advances in the Biology of Depression. Edited by Deakin JFW, Freeman H. London, Royal College of Psychiatrists, 1986, pp 71–89

Cox B, Lee TF: 5-Hydroxytryptamine induced hypothermia in rats as an in vivo model for the quantitative study of 5-hydroxytryptamine receptors. J Pharmacol Methods 5:43–51, 1981

Cox B, Lomax P: Pharmacological control of temperature regulation. Annu Rev Pharmacol Toxicol 17:341–353, 1977

Fozard JR, Mir AK, Middlemiss DN: Cardiovascular response to 8-hydroxy-2-(di-*n*-propylamino) tetralin (8-OH-DPAT) in the rat: site of action and pharmacological analysis. J Cardiovasc Pharmacol 9:328–347, 1987

Fraser W, Tucker SH, Grubb SR, et al: Effect of L-tryptophan on growth hormone and prolactin release in normal volunteers and patients with secretory pituitary tumors. Horm Metab Res 11:149–155, 1979

Fuller RW: Serotonergic stimulation of pituitary-adrenocortical function in rats. Neuroendocrinology 32:118–127, 1981

Fuller RW, Snoddy HD, Mason NR, et al: Disposition and pharmacological effects of *m*-chlorophenylpiperazine in rats. Neuropharmacology 20:155–162, 1981

Garattini S, Jori A, Buczko W, et al: The mechanism of action of fenfluramine. Postgrad Med 51 (suppl 1):27–34, 1975

Garattini S, Caccia S, Mennini T, et al: Biochemical pharmacology of the anorectic drug fenfluramine: a review. Curr Med Res Opin 6:15–27, 1979

Goa KL, Ward A: Buspirone: a preliminary review of its pharmacological properties and therapeutic efficacy as an anxiolytic. Drugs 32:114–129, 1986

Goodwin GM, Fairburn CG, Cowen PJ: The effects of dieting and weight loss on neuroendocrine responses to tryptophan, clonidine, and apomorphine in volunteers: important implications for neuroendocrine investigations in depression. Arch Gen Psychiatry 44:952–957, 1987

Green AR: 5-HT-mediated behavior. Neuropharmacology 23:1521–1528, 1984

Green AR (ed): The Neuropharmacology of Serotonin. Oxford, UK, Oxford University Press, 1985

Gudelsky GA, Koenig JI, Meltzer HY: Thermoregulatory responses to serotonin (5-HT) receptor stimulation in the rat. Neuropharmacology 25:1307–1313, 1986

Hamik A, Peroutka SJ: 1-(*m*-Chlorophenyl)piperazine (mCPP) interactions with neurotransmitter receptors in the human brain. Biol Psychiatry 25:569–575, 1989

Handwerger S, Plonk JW, Lebovitz HE, et al: Failure of 5-hydroxytryptophan to stimulate prolactin and GH secretion in man. Horm Metab Res 7:214–216, 1975

Hardy JD: Physiology of temperature regulation. Physiol Rev 41:521–606, 1961

Heuring RE, Schlegel JR, Peroutka SJ: Species variations in RU 24969 interactions with non-5-HT$_{1A}$ binding sites. Eur J Pharmacol 122:279–282, 1986

Ho BT, Schooler JC, Usdin E (eds): Serotonin in Biological Psychiatry. New York, Raven Press, 1982

Hoyer D: Functional correlates of serotonin 5-HT$_1$ recognition sites. J Recept Res 8:59–81, 1988

Hoyer D, Pazos A, Probst A, et al: Serotonin receptors in the human brain. I. Characterization and autoradiographic localization of 5-HT$_{1A}$ recognition sites: apparent absence of 5-HT$_{1B}$ recognition sites. Brain Res 376:85–96, 1986a

Hoyer D, Pazos A, Probst A, et al: Serotonin receptors in the human brain. II. Characterization and autoradiographic localization of 5-HT$_{1C}$ and 5-HT$_2$ recognition sites. Brain Res 376:97–107, 1986b

Idzikowski C, Cowen PJ, Nutt D, et al: The effects of chronic ritanserin treatment on sleep and the neuroendocrine response to L-tryptophan. Psychopharmacology (Berlin) 93:416–420, 1987

Imura H, Nakai Y, Yoshimi T: Effect of 5-hydroxytryptophan (5HTP) on growth hormone and ACTH release in man. J Clin Endocrinol Metab 36:204–206, 1973

Jacobs BL, Gelperin A (eds): Serotonin Neurotransmission and Behavior. Cambridge, MA, MIT Press, 1981

Kannengiesser MH, Hunt PF, Raynaud JP: Comparative action of fenfluramine on the uptake and release of serotonin and dopamine. Eur J Pharmacol 35:35–43, 1976

Koulu MO: Re-evaluation of L-tryptophan-stimulated human growth hormone secretion: a dose related study with a comparison with L-dopa and apomorphine tests. J Neural Transm 55:269–275, 1982

Koyama T, Lowy Mt, Meltzer HY: 5-HTP-induced cortisol response and CSF 5-HIAA in depressed patients. Am J Psychiatry 144:334–341, 1987

Krulich L, Vijayan E, Coppings RJ, et al: On the role of the central serotonergic system in the regulation of the secretion of thyrotropin and prolactin: thyrotropin-inhibiting and prolactin-releasing effects of 5-hydroxytryptamine and quipazine in the male rat. Endocrinology 105:276–283, 1979

Laakman G, Gugath M, Kuss HJ, et al: Comparison of growth hormone and prolactin stimulation induced by chlorimipramine and desipramine in man in connection with chlorimipramine metabolism. Psychopharmacology (Berlin) 82:62–67, 1984

Lamberts SWJ, MacLeod RM: The interaction of the serotonergic and dopaminergic systems on prolactin secretion in the rat. Endocrinology 103:287–295, 1978

Lancranjan I, Wirz-Justice A, Puhringer W, et al: Effect of L-5-hydroxytryptophan on growth hormone and prolactin secretion in man. J Clin Endocrinol Metab 45:588–593, 1977

Lawlor BA, Sunderland T, Hill JL, et al: Evidence for a decline with age in behavioral responsivity to the serotonin agonist, *m*-chlorophenylpiperazine, in healthy human subjects. Psychiatry Res 29:1–10, 1989

Leong DA, Frawley S, Neil JD: Neuroendocrine control of prolactin secretion. Ann Rev Physiol 45:109–127, 1983

Lewis DA, Sherman BM: Serotonergic stimulation of adrenocorticotropin in secretion of man. J Clin Endocrinol Metab 58:458–462, 1984

Lorens SA, Van de Kar LD: Differential effects of serotonin (5-HT$_{1A}$ and 5-HT$_2$) agonists and antagonists on renin and corticosterone secretion. Neuroendocrinology 45:305–310, 1987

Lowy MT, Meltzer HY: Stimulation of serum cortisol and prolactin secretion in humans by MK-212, a centrally active serotonin agonist. Biol Psychiatry 23:818–828, 1988

MacIndoe JH, Turkington RW: Stimulation of human prolactin secretion by intravenous infusion of L-tryptophan. J Clin Invest 52:1972–1978, 1973

Mann DMA, Yates PO, Marcyniuk B: Alzheimer's presenile dementia, senile dementia of Alzheimer type and Down's syndrome in middle age form an age-related continuum of pathological changes. Neuropathol Appl Neurobiol 10:185–207, 1984

McCance SL, Cowen PJ, Grahame-Smith DG: Methergoline attenuates the prolactin responses to L-tryptophan. Br J Clin Pharmacol 23:607–608, 1987

McKeith IG, Marshall EF, Ferrier IN, et al: 5-HT receptor binding in postmortem brain from patients with affective disorder. J Affective Disord 13:67–74, 1987

Meltzer HY, Lowy MT: The serotonin hypothesis of depression, in Psychopharmacology: The Third Generation of Progress. Edited by Meltzer HY. New York, Raven Press, 1987, pp 513–526

Meltzer HY, Simonovic M, Ravitz AJ: Effect of psychotomimetic drugs on rat and human prolactin and growth hormone levels, in Handbook of Psychiatry and Endocrinology. Edited by Beumont PJV, Burrows GD. New York, Elsevier Biomedical, 1982a, pp 215–238

Meltzer HY, Wiita B, Tricou BJ, et al: Effect of serotonin precursors and serotonin agonists on plasma hormone levels, in Serotonin in Biological Psychiatry. Edited by Ho BT, Schooler JC, Usdin E. New York, Raven Press, 1982b, pp 117–136

Meltzer HY, Flemming R, Robertson A: The effect of buspirone on prolactin and growth hormone secretion in man. Arch Gen Psychiatry 40:1099–1102, 1983

Meltzer HY, Lowy M, Robertson A, et al: Effect of 5-hydroxytryptophan on serum cortisol levels in major affective disorders. III. Effect of antidepressants and lithium carbonate. Arch Gen Psychiatry 41:391–397, 1984a

Meltzer HY, Perline R, Tricou BJ, et al: Effect of 5-hydroxytryptophan on serum cortisol levels in major affective disorders. II. Relation to suicide, psychosis, and depressive symptoms. Arch Gen Psychiatry 41:379–387, 1984b

Meltzer HY, Umberkoman-Wiita B, Robertson A, et al: Effect of 5-hydroxytryptophan on serum cortisol levels in major affective disorders. I. Enhanced response in depression and mania. Arch Gen Psychiatry 41:366–374, 1984c

Molliver ME: Serotonergic neuronal systems: what their anatomic organization tells us about function. J Clin Psychopharmacol 7:3S–23S, 1987

Morgan DG, May PC, Finch CE: Dopamine and serotonin systems in human and rodent brain: effects of age and neurodegenerative disease. J Am Geriatr Soc 35:334–345, 1987

Mueller EA, Murphy DL, Sunderland T: Neuroendocrine effects of m-chlorophenylpiperazine, a serotonin agonist in humans. J Clin Endocrinol Metab 61:1179–1184, 1985a

Mueller EA, Murphy DL, Sunderland T, et al: A new postsynaptic serotonin receptor agonist suitable for studies in humans. Psychopharmacol Bull 21:701–704, 1985b

Mueller EA, Murphy DL, Sunderland T: Further studies of the putative serotonin agonist m-chlorophenylpiperazine: evidence for a serotonin receptor mediated mechanism of action in humans. Psychopharmacology (Berlin) 89:388–391, 1986

Murphy DL, Campbell IC, Costa JL: The brain serotonergic system in the affective disorders. Prog Neuropsychopharmacol 2:5–31, 1978

Murphy DL, Mueller EA, Garrick NA, et al: Use of serotonergic agents in the clinical assessment of central serotonin function. J Clin Psychiatry 47 (No 4, Suppl):9–15, 1986

Murphy DL, Mueller EA, Hill JL, et al: Comparative anxiogenic, neuroendocrine, and other physiologic effects of *m*-chlorophenylpiperazine given intravenously or orally to healthy volunteers. Psychopharmacology (Berlin) 98:275–282, 1989

Murphy DL, Mueller EA, Aulakh CS, et al: Serotonin function in neuropsychiatric disorders, in Serotonin. Edited by Mylechrane EJ, de la Lande IS, Angus JA, et al. London, Macmillan (in press)

Myers RD: Serotonin and thermoregulation: old and new views. J Physiol (Paris) 77:505–513, 1981

Nakai Y, Imura H, Sakuri H, et al: Effect of cyproheptadine on human growth hormone secretion. J Clin Endocrinol Metab 3:446–449, 1974

Nutt DJ, Cowen PJ: Diazepam alters brain 5-HT function in man: implications for the acute and chronic effects of benzodiazepines. Psychol Med 17:601–607, 1987

Parati EA, Zanardi P, Cochi D, et al: Neuroendocrine effects of quipazine in man in healthy state or with neurological disorders. J Neural Transm 47:293–297, 1980

Pazos A, Palacios JM: Quantitative autoradiographic mapping of serotonin receptors in the rat brain. 1. Serotonin$_1$ receptors. Brain Res 346:205–230, 1985

Pazos A, Probst A, Palacios JM: Serotonin receptors in the human brain. III. Autoradiographic mapping of serotonin-1 receptors. Neuroscience 21:97–122, 1987

Peroutka SJ, Switzer JA, Hamik A: Identification of 5-hydroxytryptamine$_{1D}$ binding sites in human brain membranes. Synapse 3:61–66, 1989

Petraglia F, Facchinetti F, Martignoni E, et al: Serotonergic agonists increase plasma levels of β-endorphin and β-lipotropin in humans. J Clin Endocrinol Metab 59:1138–1142, 1984

Preziosi P: Serotonin control of prolactin release: an intriguing puzzle. Trends in Pharmacological Sciences 40:171–174, 1983

Quattrone A, Tedeschi G, Aguglia F, et al: Prolactin secretion in man: a useful tool to evaluate the activity of drugs on central 5-hydroxytryptaminergic neurons: studies with fenfluramine. Br J Clin Pharmacol 16:471–475, 1983

Rech RH, Gudelsky GA (eds): 5-HT Agonists as Psychoactive Drugs. Ann Arbor, MI, NPP Books, 1988

Scapagnini U, Annunziato L, DiRenzo GF: Role of the serotonergic nervous pathways on the phasic activity of the hypothalamic-hypophyseal-adrenal axis. Proc Acad Sci Med Chir 128:89–97, 1974

Siever LJ, Murphy DL, Slater S, et al: Plasma prolactin changes following fenfluramine in depressed patients compared to controls: an evaluation of central serotonergic responsivity in depression. Life Sci 34:1029–1039, 1984

Sills MA, Wolfe BB, Frazer A: Determination of selective and non-selective compounds for the 5HT₁A and 5HT₁B receptor subtypes in rat frontal cortex. J Pharmacol Exp Ther 231:480–487, 1984

Sinton CM, Fallon SL: Electrophysiological evidence for a functional differentiation between subtypes of the 5-HT₁ receptor. Eur J Pharmacol 157:173–181, 1988

Smythe GA, Gleeson RM, Stead BM: Mechanisms of 5-hydroxy-L-tryptophan-induced adrenocorticotropin release: a major role for central noradrenergic drive. Neuroendocrinology 47:389–397, 1988

Treiser SL, Cascio GS, O'Donohue TL, et al: Lithium increases serotonin release and decreases serotonin receptors in hippocampus. Science 213:1529–1531, 1981

Trulson ME: Dietary tryptophan does not alter the function of brain serotonin neurons. Life Sci 37:1067–1072, 1985

Tuomisto J, Mannisto P: Neurotransmitter regulation of anterior pituitary hormones. Pharmacol Rev 37:249–332, 1985

Van de Kar LD, Karteszi J, Bethea CL, et al: Serotonergic stimulation of prolactin and corticosterone secretion is mediated by different pathways from the mediobasal hypothalamus. Neuroendocrinology 41:380–384, 1985

van Praag HM, Lemus C, Kahn R: Hormonal probes of central serotonergic activity: do they really exist? Biol Psychiatry 22:86–98, 1987

Waeber C, Schoeffter P, Palacios JM, et al: Molecular pharmacology of 5-HT₁D recognition sites: radioligand binding studies in human, pig and calf brain membranes. Naunyn Schmiedebergs Arch Pharmacol 337:595–601, 1988

Westenberg HGM, Van Praag HM, DeJong J, et al: Postsynaptic serotonergic activity in depressive patients: evaluation of the neuroendocrine strategy. Psychiatry Res 7:361–371, 1982

Winokur A, Lindberg ND, Lucki I, et al: Hormonal and behavioral effects associated with intravenous L-tryptophan administration. Psychopharmacology (Berlin) 88:213–219, 1986

Wong DF, Wagner HN, Dannals RF, et al: Effects of age on dopamine and serotonin receptors measured by positron tomography in the living human brain. Science 226:1393–1396, 1984

Wong DF, Lever JR, Hartig PR, et al: Localization of serotonin 5-HT$_2$ receptors in living human brain by positron emission tomography using N1-([$^{11}$C]-methyl)-2-Br-LSD. Synapse 1:393–398, 1987

Wozniak KM, Aulakh CS, Hill JL, et al: The effect of 8-OH-DPAT on temperature in the rat and its modification by chronic antidepressant treatments. Pharmacol Biochem Behav 30:451–456, 1988

Yamamoto T, Hirano A: Nucleus raphe dorsalis in Alzheimer's disease: neurofibrillary tangles and loss of large neurons. Ann Neurol 17:573–577, 1985

Zohar J, Mueller EA, Insel TR, et al: Serotonin receptor sensitivity in obsessive-compulsive disorder: comparison of patients and healthy controls. Arch Gen Psychiatry 44:946–951, 1987

Zohar J, Insel TR, Zohar-Kadouch RC, et al: Serotonergic responsivity in obsessive-compulsive disorder: effects of chronic clomipramine treatment. Arch Gen Psychiatry 45:167–172, 1988

# Chapter 2

## *Developmental Influences on the Serotonergic System and Timidity in the Nonhuman Primate*

J. Dee Higley, Ph.D.
Stephen J. Suomi, Ph.D.
Markku Linnoila, M.D., Ph.D.

# Chapter 2

## Developmental Influences on the Serotonergic System and Timidity in the Nonhuman Primate

Psychiatry and psychology have a relatively long history of using nonhuman primate models to investigate various aspects of psychopathology. The use of primate models played a seminal role in initiating important areas of research such as the role of monoamines in depression (e.g., see McKinney 1988). What follows is a discussion of new data obtained using a nonhuman primate model of the relationship of developmental variables on the developing indoleamine system, with a possible link between high serotonin (5-hydroxytryptamine [5-HT]) turnover, timidity, and depression. Recently, studies with nonhuman primates have focused on the relationship between individual differences in infant temperament and subsequent behaviors characteristic of anxiety and depression. Timidity or behavioral inhibition demonstrated by a nonhuman primate infant in a novel setting is related to subsequent behavioral withdrawal, inactivity, and despair during the stress of social separation (Thompson et al. 1986). Individual differences among these infants are detectable as early as the first month of life (Thompson et al. 1987) and show continuity from infancy through childhood, adolescence, and into adulthood, and are at least partially determined by genetic influences (e.g., see Higley and Suomi 1989 for a review).

Individual differences in response to environmental novelty and/or challenge are also affected by the social environments in which the nonhuman primate subjects are reared. For example, rhesus monkeys reared from birth without adults, but with constant access to other age-mates (known as peer-only rearing), show increases in levels of fearful behavior, decreases in activity and exploration, and profound behavioral despair (Higley and Suomi, in press). Finally, when they are given free access to alcohol, peer-only-reared subjects consume more than mother-reared subjects (Dowd et al. 1988; Higley et al. 1987), suggesting that peer-only–reared subjects may be alleviating their higher levels of anxious behavior with alcohol.

Recent research with nonhuman primates has focused on the monoamine systems to explain individual differences in response to environmental challenges. Kraemer and McKinney (1979) found that pharmacologically blocking the synthesis of catecholamines with the tyrosine hydroxylase inhibitor α-methylparatyrosine (AMPT) increased levels of despair-like behavior in monkeys during the stress of social separation; moreover, a lower dose was required to produce despair in the more timid peer-only–reared monkeys than in controls reared by their mother. Subsequent studies have found that the effects of depleted monoamines on levels of despair are probably more specifically related to levels of norepinephrine. Independent of dopamine and 5-HT turnover, lower cerebrospinal fluid (CSF) concentrations of norepinephrine, whether naturally occurring or pharmacologically induced, are associated with increased despair-like behavior during social separation (Kraemer and McKinney 1979; Kraemer et al. 1984). Prior experience with social separation exacerbates behavioral withdrawal during subsequent social separations (Capitanio et al. 1986; Mineka et al. 1981; Suomi et al. 1970). Similarly, for subjects receiving previous social separations, less AMPT was needed to produce comparable levels of self-directed behavior. It is noteworthy that recent large-scale studies of human depressed patients found higher concentrations of 3-methoxy-4-hydroxyphenylglycol (MHPG) in these patients relative to nondepressed subjects (Koslow et al. 1983). Paralleling these findings in human depressed patients, Higley and colleagues (1988) found a positive correlation between concentrations of MHPG and levels of despair-like behavior during social separation. Consistent with such findings, peer-only rearing appears to be associated with increased concentrations of CSF MHPG (Higley et al. 1988).

## NONHUMAN PRIMATE STUDIES OF PERSONALITY

Recent studies in humans have linked certain psychological disorders to aberrant levels or turnover of 5-HT or its metabolites. Specifically, low levels of 5-HT have been related to increased aggression, suicide, and destructive behavior (e.g., see Åsberg et al. 1987; Mann and Stanley 1986). Links between aggression and low concentrations of 5-HT have also been found in nonhuman species (Miczek and Donat 1989). One question remaining largely unaddressed, however, concerns the opposite end of the spectrum: If diminished 5-HT activity is related to diminished inhibition of behavior, might the converse be true? Could increased 5-HT be related to over-regulation of behavior? One class of individuals who characteristically inhibit their behavior across a wide variety of situations are individuals who show personality characteristics of timidity, introversion, or shyness. Such individuals generally are less likely to approach new stimuli, are more likely to inhibit their behavior in the face of uncertainty, and take longer to warm up to new situations. As a result, the appellation recently assigned to this trait by researchers involved in its study is behavioral inhibition (Kagan et al. 1988).

Eysenck noted a negative relationship between levels of behavioral inhibition and aggression. In describing the typical extrovert and introvert, he noted the following:

> The typical extravert . . . tends to be aggressive and loses his temper quickly; altogether his feelings are not kept under control. . . . The typical introvert . . . keeps his feelings under close control, seldom behaves in an aggressive manner, and does not lose his temper easily. (Eysenck and Eysenck 1975, p. 4)

There are additional reasons to believe that timidity (behavioral inhibition) and aggression may be opposite ends of the spectrum. In rhesus monkeys, the developmental expression of each characteristic shows opposite patterns. In general, aggressive behavior is first shown in rhesus monkey behavioral repertoires in late infancy at low levels; it increases throughout the juvenile period, peaks in early adolescence, and then declines or remains constant through early and middle adulthood (Bernstein et al. 1983; Cross and Harlow 1965). Conversely, behavioral inhibition declines with age in rhesus monkeys. It occurs at high levels in early infancy, declines across childhood, and reaches and maintains its nadir by adolescence (Higley 1985; Suomi and Harlow 1976).

A similar pattern exists for sex differences in most nonhuman primate species. Levels of aggression are generally higher for males than for females, although there are major individual differences, situation and stimulus class controls, and overlap across sexes (Altmann 1968; Bernstein et al. 1983; Coelho and Bramblett 1981; Mitchell 1979; Nagel and Kummer 1974; Smuts 1987). The clearest sex differences in physical aggression appear to be in aggression directed at individuals from different groups, and in aggression that produces serious physical wounds, with males showing more of such aggression than females (Mitchell 1979; Smuts 1987). While approach and exploratory behavior has been less investigated than aggression, when sex differences in regard to the former behavior are observed, males are typically more likely to approach novel objects and spend more time in exploration than females (Mitchell 1979; Stevenson-Hinde and Zunz 1978).

For rhesus monkeys, early rearing conditions, such as peer-only rearing with more than two partners, affect the expression of aggression and behavioral inhibition in opposite directions, decreasing levels of aggression and increasing levels of behavioral inhibition (e.g., see Chamove et al. 1973; Coelho and Bramblett 1981; Harlow and Harlow 1969; see also Suomi 1982 for an exception under some conditions). It is noteworthy that in humans, both behavioral inhibition and aggression stabilize early in life and show continuity across developmental periods, a pattern similar to that in nonhuman primates. Indeed, these two traits may be some of the earliest to show continuity with significant correlations across decades in humans (e.g., see Moss and Susman 1980 for a recent review; see also Eron 1987; Huesmann et al. 1984; Kagan et al. 1988).

Aggression may even be transmitted across generations. When children were asked who the most aggressive person in their class was, the boys whom the class rated as bullies were likely to be the offspring of fathers who two decades earlier were rated as the most aggressive boys (Eron 1987; Huesmann et al. 1984). Both timidity and aggression appear to be influenced by genetics (e.g., see Buss and Plomin 1984; Daniels and Plomin 1985; Fuller and Thompson 1978; Ghodsian-Carpey and Baker 1987; Mednick et al. 1984; Scarr and Kidd 1983).

If disorders of impulsivity are related to low central nervous system (CNS) 5-HT turnover, behaviors characteristic of over-inhibition might be related to high 5-HT turnover. There is evidence for this proposal. Studies have shown that changes in the indoleamine system may parallel developmental patterns and sex-related differences in aggressive behaviors. Younger subjects show higher concentrations of 5-HT or its metabolites, and males, particularly in depressed populations, tend to have lower concentrations than females (Andersson and Roos 1969; Bowers and Gerbode 1968; Koslow et al. 1983; Post et al. 1980; Rogers and Dubowitz 1970; Silverstein et al. 1985). Similar to the pattern seen in behavioral inhibition and aggression, there is some evidence that concentrations of 5-HT metabolites show continuity across time (Åsberg et al. 1987; Coppen et al. 1972; Träskman-Bendz et al. 1984; van Praag and Korf 1971; Yuwiler et al. 1981), although unlike behavioral inhibition, continuity of 5-HT metabolites has only been investigated in adults. Like behavioral inhibition, the rate of 5-HT turnover may be heritable, although the extent of heritability is not precisely defined (Agras 1985; Fuller and Thompson 1978; Scarr and Kidd 1983; Sedvall et al. 1980, 1984). Studies in both humans and monkeys indicate that behavioral inhibition is affected by rearing history (Daniels and Plomin 1985; Higley and Suomi 1989). However, whether 5-HT activity is affected by rearing history has not been systematically investigated in primates.

In a series of investigations conducted over a 3-year period, our group studied the relationship between CSF 5-hydroxyindoleacetic acid (5-HIAA) concentration and behavioral timidity, focusing on how 5-HT turnover may be related to coping with environmental stress. The stressors utilized in these studies were novelty and chronic repeated social separation. Six general hypotheses were investigated in two studies. First, previous research has indicated that the relative incidence of aggression increases and that of timidity decreases with age; hence, we hypothesized that CSF 5-HIAA concentration would decrease with age in rhesus monkeys. Second, male rhesus monkeys generally appear to be less inhibited behaviorally than females; conversely, males generally show more aggression than females. Therefore, CSF 5-HIAA concentrations should be lower in males. Third, rearing conditions that are associated with relatively low levels of aggression and high levels of behavioral inhibition, such as peer-only rearing, should also be associated with relatively high concentrations of CSF 5-HIAA. Fourth, because individual differences in behavioral expressions of aggression and timidity show continuity across time, individual differences in concentrations of CSF 5-HIAA should also show continuity across time. Fifth, in terms of individual differences in behavior, high CSF 5-HIAA concentrations should correlate positively with levels of behavioral inhibition. Finally, CSF 5-HIAA concentrations have been shown to be low in a subgroup of human depressed patients. If CSF 5-HIAA concentration correlates with levels of behavioral inhibition, we hypothesized that behavioral expressions of timidity would show a positive correlation with subsequent despair-like behaviors during separation.

## STUDY 1: GENERAL EXPERIMENTAL PROCEDURES

To test the hypotheses discussed above, CSF was repeatedly obtained from a total of 37 rhesus monkeys when they were 6 months of age (year 1), either while they were in their social groups or while they were separated from their attachment sources. Data were available for 20 of these same subjects, under conditions identical to those of year 1, when they were 18 months of age (year 2). Subjects were reared either in mother-infant dyads ($n = 13$) or in peer-only groups

($n = 12$).    The peer-only subjects were removed from their mothers at birth and raised for 6 months without adults but with constant access to three other age-mates. Subsequent to the 6-month separations, all 20 of the remaining subjects were maintained under identical conditions. This was achieved by placing each group of peer-only–reared subjects with a group of the age-matched mother-reared counterparts. During both years, subjects underwent four sequential 4-day social separations, each followed by 3 days of home-cage reunions. The separations were performed by placing the subjects in single cages in a novel room where they could hear, but not touch or see, the other separated infants. CSF was obtained from the cisterna magna three times for each subject: prior to separation, during the first week, and during the fourth week of separation. All results described in the following are significant at the $P < .05$ level of probability.

### Early Rearing and Sex Differences

The results were generally as predicted. As discussed earlier, only a few nonhuman primate investigations exist that focus on sex differences in timidity; but in general they indicate that females show greater behavioral inhibition than males (Mitchell 1979). On the other hand, levels of aggression show a more clear-cut sex difference, with males exhibiting higher levels. Consistent with our predictions, female mother-reared subjects showed significantly higher concentrations of 5-HIAA than male mother-reared subjects and both male and female peer-only–reared subjects (see Figure 2-1).

These findings indicate that young female monkeys show increased 5-HT turnover compared to males. In human populations this relationship has been reported only in adults. Future studies will investigate whether, as in humans, these sex differences are retained into adulthood. The present findings suggest that if 5-HT is influencing the expression of behavioral inhibition and aggression in rhesus monkeys, then pharmacologically decreasing 5-HT may have differential effects on aggression and timidity according to sex. To date, studies have not directly addressed this possibility.

### Changes Across Years in Cerebrospinal Fluid 5-HIAA Concentration and Group Effects

Comparisons of the 6- and 18-month separation data indicated a decline in concentrations of 5-HIAA from the 6- to 18-month separations; however, peer-only–reared subjects demonstrated less of a decline, maintaining significantly higher concentrations at the 18-month time point (see Figure 2-2). Thus, as hypothesized, CSF 5-HIAA concentrations paralleled the decline with age seen in

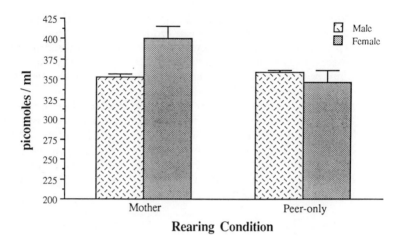

**Figure 2-1.** Levels of cisternal CSF 5-HIAA for males and females as a result of being reared either with their mother or in a peer-only condition; female mother-reared subjects had significantly higher levels of CSF 5-HIAA than did subjects reared under any other condition ($P < .05$).

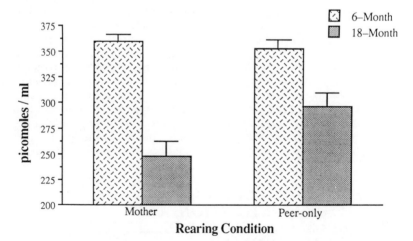

**Figure 2-2.** Levels of cisternal CSF 5-HIAA for mother- and peer-only–reared subjects at 6 and 18 months of age. Six-month-old subjects had significantly higher levels of CSF 5-HIAA than did 18-month-old subjects, but levels in peer-only–reared subjects declined less ($P < .05$).

humans, in which the highest concentrations of 5-HIAA in lumbar CSF are found in infancy, with an age-related decline thereafter. Also as predicted, peer-only rearing, a condition that increases timidity and decreases aggression, was associated with expected increases in concentrations of 5-HIAA. Peer-only–reared subjects showed higher concentrations of 5-HIAA than mother-reared subjects. It is worth noting that during the stress of social separation, peer-only rearing tended to increase the incidence of despair-like depressive behaviors. This finding suggests a relationship between behavioral inhibition, 5-HIAA, and depressive behaviors. Whether similar parallels between 5-HIAA and depression-like behavior exist for sex differences is largely unknown (although this may simply be due to researchers not directly analyzing their data for sex differences).

## Continuity

There were strong correlations between the CSF 5-HIAA concentrations from year 1 to year 2. The mean concentrations of 5-HIAA obtained at 6 months of age correlated with the mean concentrations at 18 months of age ($r = .51$). Although some researchers have found that in adult male monkeys whole-blood 5-HT levels are stable over short periods of time (Raleigh et al. 1984), this is to our knowledge the first indication of an index of CNS 5-HT turnover showing temporal continuity between infancy and childhood. We noted above that among adult human populations individual differences in CSF 5-HIAA concentrations showed stable group rankings across time in both control subjects and depressed patients. This was true even though CSF 5-HIAA concentrations in the depressed patients increased subsequent to resolution of the depressive episode (Träskman-Bendz et al. 1984). Such findings are an indication that the activity of the indoleamine system may vary in a trait-like fashion across individuals, stabilizing early in life and showing continuity over time. Subsequent studies are already underway to follow the present rhesus monkey subjects into adulthood to assess the long-term stability of the serotonergic systems.

## STUDY 2: NEONATAL TEMPERAMENT, BEHAVIORAL INHIBITION, AND CEREBROSPINAL FLUID 5-HIAA CONCENTRATION

Eleven subjects raised in peer-only groups participated in a second study. Data concerning neonatal temperament (fearfulness and consolability) were obtained on these 11 subjects during the first month when they were in the neonatal nursery (see Schneider 1987 for definitions and procedures). Additional data concerning behavioral inhibition were obtained on the subjects when they were 4 months old during exposure to a novel environment. This environment was a playroom with ladders, toys, and other novel objects, and included an age-matched unfamiliar monkey. A total inhibition score was obtained by taking an aggregate average z score for each subject based on (a) the latency to touch each object, (b) the latency to enter new quadrants, and (c) the total time orienting to the unfamiliar monkey as the unfamiliar monkey explored the playroom. Consistent with our first study (see above), CSF was obtained from these subjects when they were 6 months old during a series of four 4-day social separations, identical to those previously described.

## Correlation of 5-HIAA Concentration With Behavior

There was a significant correlation between ratings of fearfulness during the first month and mean 5-HIAA concentrations during the separation at 6 months ($r$ = .847; see Figure 2-3). For the exposure to the novel room at 4 months, when the behaviors related to timidity were summarized in a mean behavioral inhibition score for each subject, there was a significant correlation between the behavioral inhibition score and 5-HIAA concentrations during the fourth week of the separation at 6 months ($r$ = .650).

When the mean score for behavioral inhibition for each individual

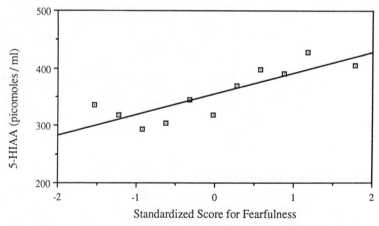

**Figure 2-3.** Correlation among standardized fearfulness ratings taken during the first month of life and mean levels of CSF 5-HIAA taken during social separations at 6 months of life ($r$ = .847, $n$ = 11, $P$ < .01).

was correlated with mean levels of self-clasping displayed during the social separation at 6 months (a primary measure used to rate despair-like behaviors in monkeys), there was a significant positive correlation between levels of behavioral inhibition and the despair behavior of self-clasping observed during both the early acute and later chronic phases of social separation ($r$ = .742 and .701, respectively).

Thus our second study revealed a positive correlation between individual differences in behavioral inhibition in early infancy, as measured by neonatal temperament ratings, and subsequent concentrations of CSF 5-HIAA. The correlation was still present 4 months later when behavioral inhibition was assessed using an objective scoring paradigm in structured playroom tests. The correlation was evident despite major developmental changes in behavior between the neonatal period and middle infancy, and despite obvious differences in behavioral paradigms in the two test settings. If 5-HT is indeed mediating behavioral inhibition, our findings indicate that these individual differences could be present very early in life.

The sample size in this study was small; thus the results should be replicated. Future studies should also obtain CSF samples early in life to assess the youngest age at which individual differences in 5-HIAA can be measured, and when they stabilize. The positive correlations between behavioral inhibition and CSF 5-HIAA concentration indicate that this may also be a promising area of research for understanding disorders of overinhibition. Studies are already underway with these monkey subjects to assess the relationship of early individual differences in 5-HT activity and other possibly related behavioral disorders, such as excessive consumption of alcohol and maltreatment of offspring as these subjects approach adulthood.

Finally, our findings indicate that early expressions of behavioral inhibition may be a behavioral marker or risk factor for subsequent depression-like behaviors in the face of separation stress. While extreme timidity has largely been seen as a normal variation of behavior in approximately 10 to 15% of human children (Kagan et al. 1988), our findings suggest that these individuals may be at risk for future problems with anxiety and depression if they are exposed to stressful life events.

## RELATIONSHIP BETWEEN CENTRAL SEROTONERGIC ACTIVITY, DEPRESSION, AND TIMIDITY/AGGRESSION

One of the paramount problems in studying primary affective disorders is identifying which individuals are at risk for developing depressive symptomatology and what traits or markers could be used to identify these individuals. Among human clinical populations, one of the earliest postulated biological mechanisms for depression was diminished CNS 5-HT. Two decades later 5-HT still occupies a central place in biochemical theories of depression (Meltzer and Lowy 1987); however, theories of depression have moved from global hypotheses such as depression is caused by insufficient levels of some specific neurotransmitter, to more specific theories that link individual symptoms or types of depression to specific abnormalities or imbalances of neurotransmitter systems.

Studies of nonhuman primates shed further light on the relationship between 5-HT and one possible form of depression-like behavior. Our findings suggest a positive relationship between expressions of timidity early in life and 5-HT metabolism. Behavioral inhibition may serve as a behavioral marker, and indices of increased 5-HT turnover as a biological marker, to identify individuals at risk for displaying depressive responses to separation. Indeed, these may be factors predisposing individuals to depression. Peer-only rearing increases timidity and concentrations of CSF 5-HIAA; when these monkeys are subjected to social separation, they are more likely to show despair-like behaviors. In addition, within-group individual differences in behavioral inhibition are significantly correlated with emergence and severity of depressive behaviors during the stress of social separation. Considered together, these findings indicate that behavioral inhibition early in life and high concentrations of CSF 5-HIAA may be interdependent risk factors for subsequent depression.

This postulated positive correlation between 5-HT and depression is contrary to early human studies that postulated globally diminished levels of 5-HT in depression. However, while in some studies the average concentration of CSF 5-HIAA in depressed patients has been shown to be lower than in nondepressed control subjects, this may be due to the inclusion of a subgroup of depressed patients who are prone to self-injuring behaviors such as suicide (e.g., see Åsberg et al. 1987). While some investigations have found CSF 5-HIAA levels to be lower in depressed patients than in nondepressed control subjects, more recent data have found higher mean levels, suggesting that overactive 5-HT mechanisms may also be related to depression (Koslow et al. 1983; Ogren et al. 1979). Our findings are in general agreement with these later proposals.

## ANXIETY

High levels of anxiety have been linked to increased 5-HT activity (Hoehn-Saric 1982; Stein 1981). Recent studies using young rhesus monkeys indicate that subjects who show high levels of despair during a long-term social separation also show high levels of anxious-like behavior (Higley 1985; Higley and Suomi 1989). Indeed, the strongest predictor of despair during social separation is high anxiety both prior to and during the initial phase of separation (Higley 1985). In humans, anxiety is a common symptom in major depressions (American Psychiatric Association 1987), but it is more severe in some patients than in others (Agras 1985; Akiskal 1984; Van Valkenburg et al. 1984). Some investigations suggest that perhaps there is a subgroup of depressed individuals with chronically high levels of anxiety (Akiskal 1984; Gersh and Fowles 1979; Gray 1985). Gray (1985) proposes that GABA-ergic inhibition of the major serotonergic cells of the CNS, the raphe nuclei, may underlie the action of many antianxiety drugs. Our findings of high timidity and fearfulness in subjects who are prone to despair, with concurrent increased 5-HT turnover, in conjunction with these other studies, support the speculation that perhaps a subgroup of depressed individuals who show high levels of anxiety may have higher than usual levels of 5-HT activity.

One difficulty researchers working with clinical populations face is that different studies find lower, higher, or similar levels of CSF 5-HT or its metabolites in different studies of depressed patients (Meltzer and Lowy 1987). The choice of different depressed patient subgroups with different symptom profiles may account for these differences across studies in terms of CSF 5-HIAA concentrations. When depressed patients with anxiety are pooled with depressed individuals who show suicidal behavior (a group who have shown lower levels of 5-HIAA), each subgroup may cancel out each other's impact, resulting in pooled mean levels similar to those of control subjects. Depending on the selection of subjects from different pools (suicidal vs. timid-anxious), mean levels of 5-HIAA could be low, normal, or high.

In discussing the results, a caveat is in order. It should be noted that our results are correlational. Thus, cause and effect are difficult to establish. Future studies should selectively manipulate CNS 5-HT activity and observe the effects on behavioral inhibition and stress-induced despair in monkeys. There is, however, evidence that at least in terms of aggression, pharmacologically depleting 5-HT using $p$-chlorophenylalanine increases levels of aggression in a related species, the vervet monkey (*Cercopithecus aethiops sabaeus*) (Raleigh 1987; Raleigh et al. 1980). Pharmacological tests to explore 5-HT's role in the behavioral expression of timidity have not to our knowledge been conducted in monkeys.

While we focused primarily on possible 5-HT mechanisms underlying behavioral expressions of timidity, aggression, and depression, other neurotransmitter systems are undoubtedly involved. Kagan found a relationship between noradrenergic activity and behavioral inhibition (Kagan et al. 1988). We also found that relative to mother-reared subjects, peer-only–reared subjects show increased concentrations of CSF MHPG following social separation (Higley et al. 1988). Furthermore, even when group differences are factored out, there is still a correlation between individual differences in levels of MHPG and despair-like behavior during separation (Higley et al. 1988). The role of neurotransmitter interactions in these processes will be examined in future studies.

## CONCLUSIONS

Developmental, rearing, and gender differences in the serotonergic system, as characterized by CSF 5-HIAA concentration measurements, appear to parallel previously established behavioral differences in behavioral measures of timidity and aggression. When behavior–5-HIAA relationships are directly assessed, concentrations of CSF 5-HIAA correlate directly with measures of behavioral inhibition. In addition, behavioral inhibition appears to correlate with levels of despair observed during social separation.

Accordingly, examination of indices of 5-HT activity suggests a relationship between 5-HT, timidity, and depressive behaviors. While, in general, previous studies have investigated the relationship between impulsive aggression, or other disorders of impulse control, and low concentrations of CSF 5-HIAA, these results indicate that the opposite end of the spectrum is also worthy of consideration. CNS 5-HT activity may be related to overall inhibition of behavior (Soubrie 1986), with low levels being related to underinhibition, and high levels related to overinhibition of behavior, specifically timidity in the young rhesus monkey. Children showing high levels of timidity may also be at risk for subsequent psychopathology. This may be in part related to overactivity of CNS serotonergic systems.

## REFERENCES

Agras WS: Panic. New York, WH Freeman, 1985

Akiskal HS: The interface of chronic depression with personality and anxiety disorders. Psychopharmacol Bull 20:393–398, 1984

Altmann SA: Sociobiology of rhesus monkeys. IV. Testing Mason's hypothesis of sex differences in affective behavior. Behaviour 32:49–69, 1968

American Psychiatric Association: Diagnostic and Statistical Manual of Mental Disorders, 3rd Edition, Revised. Washington, DC, American Psychiatric Association, 1987

Andersson H, Roos BE: 5-Hydroxindoleacetic acid in cerebrospinal fluid of hydrocephalic children. Acta Paediatr Scand 58:601–608, 1969

Åsberg M, Schalling D, Träskman-Bendz L, et al: Psychobiology of suicide, impulsivity, and related phenomena, in Psychopharmacology: The Third Generation of Progress. Edited by Meltzer HY. New York, Raven Press, 1987, pp 655–668

Bernstein I, Williams L, Ramsay M: The expression of aggression in old world monkeys. Int Primatol 4:113–124, 1983

Bowers MB, Gerbode FA: Relationship of monamine metabolites in human cerebrospinal fluid to age. Nature 219:156–157, 1968

Buss AH, Plomin R: Temperament: Early Developing Personality Traits. Hillsdale, NJ, Erlbaum, 1984

Capitanio JP, Rasmussen KLR, Snyder DS, et al: Long-term follow-up of previously separated pigtail macaques: group and individual differences in response to novel situations. J Child Psychol 27:531–538, 1986

Chamove AS, Rosenblum LA, Harlow HF: Monkeys (*Macaca mulatta*) raised only with peers: a pilot study. Anim Behav 21:316–325, 1973

Coelho AM, Bramblett CA: Effects of rearing on aggression and subordination in papio monkeys. Am J Primatol 1:401–412, 1981

Coppen A, Prange AJ, Whybrow PC, et al: Abnormalities of indoleamines in affective disorders. Arch Gen Psychiatry 26:474–478, 1972

Cross HA, Harlow HF: Prolonged and progressive effects of partial isolation on the behavior of macaque monkeys. Journal of Experimental Research in Personality 1:39–49, 1965

Daniels D, Plomin R: Origins of individual differences in infant shyness. Dev Psychol 21:118–121, 1985

Dowd BM, Higley JD, Suomi SJ, et al: Alcohol consumption as a function of social separation, individual differences in despair, and early rearing experiences in rhesus monkeys. Am J Primatol 14:417–418, 1988

Eron LD: The development of aggressive behavior from the perspective of a developing behaviorism. Am Psychol 42:435–442, 1987

Eysenck HJ, Eysenck SBG: Manual of the Eysenck Personality Questionnaire. San Diego, CA, Educational and Industrial Testing Service, 1975

Fuller JL, Thompson WR: Foundations of Behavior Genetics. St. Louis, MO, CV Mosby, 1978

Gersh FS, Fowles DC: Neurotic depression: the concept of anxious depression, in The Psychobiology of the Depressive Disorders. Edited by Depue RA. New York, Academic Press, 1979, pp 81–104

Ghodsian-Carpey J, Baker LA: Genetic and environmental influences on aggression in 4- to 7-year-old twins. Aggressive Behav 13:173–186, 1987

Gray JA: Issues in the neuropsychology of anxiety, in Anxiety and the Anxiety Disorders. Edited by Tuma AH, Maser J. Hillsdale, NJ, Erlbaum, 1985, pp 5–25

Harlow HF, Harlow MK: Effects of various mother-infant relationships on rhesus monkey behaviors, in Determinants of Infant Behavior, Vol 4. Edited by Foss BM. London, Methuen, 1969, pp 15–36

Higley JD: Continuity of social separation behaviors in rhesus monkeys from infancy to childhood. Unpublished doctoral dissertation, University of Wisconsin-Madison, 1985

Higley JD, Suomi SJ: Temperamental reactivity in nonhuman primates, in Handbook of Temperament in Childhood. Edited by Kohnstamm D, Bates JE, Rothbart MK. New York, John Wiley, 1989, pp 153–168

Higley JD, Linnoila M, Suomi SJ, et al: Early peer-only rearing increases ethanol consumption in rhesus monkeys (*Macaca mulatta*). Am J Primatol 12:348, 1987

Higley JD, Suomi SJ, Linnoila M: Central amine correlates of timidity and affective disturbances in rhesus monkeys. Am J Primatol 14:425, 1988

Hoehn-Saric R: Neurotransmitters in anxiety. Arch Gen Psychiatry 39:735–742, 1982

Huesmann LR, Eron LD, Lefkowitz MM, et al: Stability of aggression over time and generations. Dev Psychol 20:1120–1134, 1984

Kagan J, Reznick JS, Snidman N: Biological bases of childhood shyness. Science 240:167–171, 1988

Koslow SH, Maas JW, Bowden CL, et al: CSF and urinary biogenic amines and metabolites in depression and mania: a controlled, univariate analysis. Arch Gen Psychiatry 40:999–1010, 1983

Kraemer GW, McKinney WT: Interactions of pharmacological agents which alter biogenic amine metabolism and depression. J Affective Disord 1:33–54, 1979

Kraemer GW, Ebert MH, Lake CR, et al: Cerebrospinal fluid measures of neurotransmitter changes associated with pharmacological alteration of the despair response to social separation in rhesus monkeys. Psychiatry Res 11:303–315, 1984

Mann JJ, Stanley M (eds): Psychobiology of Suicidal Behavior (special issue). Ann NY Acad Sci, Vol 487, 1986

McKinney WT: Models of Mental Disorders: A New Comparative Psychiatry. New York, Plenum, 1988

Mednick SA, Gabrielli WF, Hutchings B: Genetic influences in criminal convictions: evidence from an adoption court. Science 224:891–893, 1984

Meltzer HY, Lowy M: The serotonin hypothesis of depression, in Psychopharmacology: The Third Generation of Progress. Edited by Meltzer HY. New York, Raven Press, 1987, pp 513–526

Miczek KA, Donat P: Brain 5-HT systems and inhibition of aggressive behavior, in Behavioral Pharmacology of 5-HT. Edited by Archer T, Bevan P, Cools A. Hillsdale, NJ, Erlbaum, 1989, pp 117–144

Mineka S, Suomi SJ, DeLizio R: Multiple separations in adolescent monkeys: an opponent-process interpretation. J Exp Psychol [Gen] 110:56–85, 1981

Mitchell G (ed): Behavioral Sex Differences in Nonhuman Primates. New York, Van Nostrand Reinhold, 1979

Moss HA, Susman EJ: Longitudinal study of personality development, in Constancy and Change in Human Development. Edited by Brim OG, Kagan J. London, Harvard University Press, 1980, pp 530–595

Nagel U, Kummer H: Variation in cercopithecoid aggressive behavior, in Primate Aggression, Territoriality, and Xenophobia. Edited by Holloway RL. New York, Academic Press, 1974, pp 159–184

Ogren SO, Fuxe K, Agnati LF, et al: Reevaluation of the indoleamine hypothesis of depression. Evidence for a reduction of functional activity of central 5-HT systems by antidepressant drugs. J Neural Transm 46:85–103, 1979

Post RM, Ballenger JC, Goodwin FK: Cerebrospinal fluid studies of neurotransmitter function in manic and depressive illness, in Neurobiology of Cerebrospinal Fluid I. Edited by Wood JH. New York, Plenum, 1983, pp 685–717

Raleigh MJ: Differential behavioral effects of tryptophan and 5-hydroxytryptophan in vervet monkeys: influence of catecholaminergic systems. Psychopharmacology (Berlin) 93:44–50, 1987

Raleigh MJ, Brammer GL, Yuwiler A, et al: Serotonergic influences on the social behavior of vervet monkeys (*Cercopithecus aethiops sabaeus*). Exp Neurol 68:322–334, 1980

Raleigh MJ, McGuire MT, Brammer GL, et al: Social and environmental influences on blood serotonin concentrations in monkeys. Arch Gen Psychiatry 41:405–410, 1984

Rogers JJ, Dubowitz V: 5-Hydroxindoles in hydrocephalus: a comparative study of cerebrospinal fluid and blood levels. Dev Med Child Neurol 11:461–466, 1970

Scarr S, Kidd KK: Developmental behavior genetics, in Handbook of Child Psychology, Vol 2. Infancy and Developmental Psychobiology. Edited by Mussen PH. New York, John Wiley, 1983, pp 346–443

Schneider ML: A rhesus monkey model of human infant individual differences. Unpublished doctoral dissertation, University of Wisconsin, Madison, WI, 1987

Sedvall G, Fyro B, Gullberg B, et al: Relationships in healthy volunteers between concentrations of monoamine metabolites in cerebrospinal fluid and family history of psychiatric morbidity. Br J Psychiatry 136:366–374, 1980

Sedvall G, Iselius L, Nyback H, et al: Genetic studies of CSF monoamine metabolites, in Frontiers in Biochemical and Pharmacological Research in Depression. Edited by Usdin E, Åsberg M, Bertilsson L, et al. New York, Raven Press, 1984, pp 79–85

Silverstein FS, Johnston MV, Hutchinson RJ, et al: Lesch-Nyhan syndrome: CSF neurotransmitter abnormalities. Neurology 35:907–911, 1985

Smuts BB: Gender, aggression, and influence, in Primate Societies. Edited by Smuts BB, Cheney DL, Seyfarth RM, et al. Chicago, IL, University of Chicago Press, 1987, pp 400–412

Soubrie P: Reconciling the role of central serotonin neurons in human and animal behavior. Behav Brain Sci 9:319–364, 1986

Stein L: Behavioral pharmacology of benzodiazepines, in Anxiety: New Research and Changing Concepts. Edited by Klein DF, Rabkin J, New York, Raven Press, 1981, pp 201–213

Stevenson-Hinde J, Zunz M: Subjective assessment of individual rhesus monkeys. Primates 19:473–482, 1978

Suomi SJ: Abnormal behavior in nonhuman primates, in Primate Behavior. Edited by Fobes J, King J. New York, Academic Press, 1982, pp 171–215

Suomi SJ, Harlow HF: The facts and functions of fear, in Emotion and

Anxiety. Edited by Zuckerman M, Spielberger CD. Hillsdale, NJ, Erlbaum, 1976, pp 3–34

Suomi SJ, Harlow HF, Domek CJ: Effect of repetitive infant-infant separation of young monkeys. J Abnorm Psychol 76:161–172, 1970

Thompson WW, Higley JD, Byrne EA, et al: Behavioral inhibition in nonhuman primates: psychobiological correlates and continuity over time. Paper presented at the meeting of the International Society for Developmental Psychobiology, Annapolis, MD, November 1986

Thompson WW, Schneider ML, Higley JD, et al: Early temperament as a predictor of subsequent heart rate in rhesus monkeys. Am J Primatol 12:374, 1987

Träskman-Bendz L, Åsberg M, Bertilsson L, et al: CSF monoamine metabolites of depressed patients during illness and after recovery. Acta Psychiatr Scand 69:333–342, 1984

van Praag HM, Korf J: Endogenous depressions with and without disturbances in the 5-hydroxytryptamine metabolism: a biochemical classification? Psychopharmacology (Berlin) 19:148–152, 1971

Van Valkenburg C, Akiskal HS, Puzantian V, et al: Clinical, family history, and naturalistic outcome—comparisons with panic and major depressive disorders. J Affective Disord 6:67–82, 1984

Yuwiler A, Brammer GL, Morley JE, et al: Short-term and repetitive administration of oral tryptophan in normal men. Arch Gen Psychiatry 38:619–626, 1981

# Chapter 3

## *Serotonin in Autism*

P. Anne McBride, M.D.
George M. Anderson, Ph.D.
J. John Mann, M.D.

# Chapter 3

## *Serotonin in Autism*

Autistic disorder is a syndrome that presents in early childhood and is characterized by prominent distortions in social, language, and cognitive development (American Psychiatric Association 1987; Ornitz and Ritvo 1976; Rutter 1985). Approximately two to five children per 10,000 live births suffer from the disorder, which is three to four times more common in males than in females. Impaired capacity for social relationships is an essential feature of autistic disorder. Autistic youngsters show a pervasive lack of interest in other people and in social or imaginative play. Most autistic children are delayed in the acquisition of both verbal and nonverbal communication skills, and many never develop useful language. Autistic children with language typically exhibit deviant speech patterns including echolalia and pronominal reversal (for instance, substitution of "he" for "I"). In addition, most autistic youngsters exhibit motor stereotypies, such as twirling and flapping of the hands. Self-abusive behaviors are common. Autistic children are characteristically resistant to change and are prone to engage in repetitive behaviors or rituals. While some autistic individuals exhibit increased interest in social relationships and fewer bizarre behaviors as they enter adolescence or young adulthood, virtually all continue to have moderate to severe impairment in psychosocial functioning. DSM-III-R criteria (APA 1987) for autistic disorder, which are more broadly defined than DSM-III criteria (APA 1980) for infantile autism, allow for the fact that symptomatology may change with increasing maturity.

Portions of this chapter were presented at the annual meeting of the American Psychiatric Association, May 1987, Chicago, Illinois, and appeared in similar form in the *Archives of General Psychiatry*, Vol. 46, 1989, pp. 213–221. Research conducted in our laboratories was supported by the Stallone Fund for Autism Research; Biomedical Research Support Grant No. SO7PR05396, National Institutes of Health; and National Institutes of Health Grant No. HM-30929. Dr. McBride is the recipient of a Teacher-Scientist Award from the Andrew W. Mellon Foundation and a Reader's Digest Research Fellowship. Dr. Mann is the recipient of an Irma T. Hirschl Career Scientist Award.

In the current era, it is generally accepted that dysfunction of the central nervous system (CNS) is implicated in the pathogenesis of autism (Ornitz and Ritvo 1976; Rutter 1985). Three-quarters of autistic children have mild to severe mental retardation, and a quarter develop seizures during later childhood or adolescence. The autistic syndrome is frequently present in children with congenital rubella, phenylketonuria (PKU), and chromosomal abnormalities such as fragile X (Brown et al. 1982). While studies employing computed tomography have failed to reveal a specific defect in brain structure in autism (Prior et al. 1984; Rosenbloom et al. 1984), hypoplasia of a portion of the cerebellar vermis was found in a majority of 18 higher-functioning autistic children and young adults recently evaluated by magnetic resonance imaging (Courchesne et al. 1988). Regardless of whether a neuroanatomical defect is demonstrated in autism, alterations in CNS function are most likely present at the cellular level, with involvement of neurotransmitter systems.

Alterations in indices of function of the neurotransmitter serotonin (5-hydroxytryptamine [5-HT]) have been reported in persons with affective disorders (van Praag 1986), suicidal behavior (Mann et al. 1986; van Praag 1986), a history of impulsivity or aggression (Siever et al. 1986), mental retardation (Hanley et al. 1977; Lott et al. 1972), or Alzheimer's disease (Morgan et al. 1987). Since disturbances of affect or mood (e.g., severely constricted affect, emotional lability), mental retardation, self-injurious behavior, and temper tantrums are common in autistic individuals, alterations in serotonergic function have been sought in autism. This chapter will review evidence that suggests that serotonergic mechanisms may indeed be altered in at least some persons with autistic disorder. Data from our recently published pilot study (McBride et al. 1989) of serotonergically mediated physiological responses in autistic young adults will be presented in more detail.

## PSYCHOBIOLOGIC STUDIES OF SEROTONIN IN AUTISM

### Blood Serotonin Content

In 1961, Schain and Freedman reported the presence of elevated levels of 5-HT in the whole blood of 6 of 23 autistic children. Since then, a number of studies (Anderson et al. 1987; Hanley et al. 1977; Ritvo et al. 1970; Takahashi et al. 1976) have confirmed that mean concentrations of 5-HT in blood are significantly higher in groups of autistic subjects than in groups of age- and gender-matched normal control subjects. Most recently, Anderson et al. (1987) found a mean whole-blood 5-HT concentration of 205 ± 16 ng/ml (± SEM) in a sample of 21 drug-free autistic subjects vs. 136 ± 5.4 ng/ml in a control group consisting of 87 normal volunteers. While most studies have indicated that approximately one-third of autistic subjects have blood 5-HT levels exceeding those typically measured in normal control subjects (Hanley et al. 1977; Ritvo et al. 1970; Schain and Freedman 1961), the Gaussian distribution of values suggests that those subjects with higher levels do not constitute a distinct subgroup of autistic individuals (Anderson et al. 1987). Correlations have generally not been found between blood 5-HT content and specific autistic behaviors (Anderson et al. 1987; Hanley et al. 1977; Volkmar et al. 1983a) or the presence of neurological stigmata (Hanley et al. 1977), although some studies have suggested an association between hyperserotonemia and low IQ in autistic children (Campbell et al. 1975; Hanley et al. 1977).

The finding of elevated whole-blood 5-HT content is not specific to autistic disorder. Hyperserotonemia is present in medical disorders, most notably carcinoid syndrome (Crawford et al. 1967), and has been reported in other neuropsychiatric disorders, including schizophrenia (DeLisi et al. 1981). Approximately one-half of severely retarded children (excluding children with Down's syndrome or PKU) without prominent autistic symptoms have elevated blood 5-HT levels (Hanley et al. 1977; Tu and Partington 1972). On the other hand, hyperserotonemia is not found in all disorders in which mental retardation is a characteristic feature. Blood 5-HT content is substantially reduced in individuals with Down's syndrome (Lott et al. 1972; Tu and Partington 1972) or PKU (Tu and Partington 1972).

It is not known whether blood 5-HT content bears any relationship to serotonergic activity in the CNS. Blood 5-HT is largely synthesized by the enterochromaffin cells of the intestinal tract. Almost all 5-HT in blood is sequestered in cytoplasmic storage granules within the blood platelet (Sneddon 1973). Peripheral 5-HT is catabolized by monoamine oxidase (MAO) primarily in the lung and liver (Thomas and Vane 1967).

Efforts to explain the presence of altered blood 5-HT content in autism and retardation have focused primarily on the evaluation of platelet 5-HT uptake and storage. In the case of Down's syndrome, hyposerotonemia is thought to result from decreased active transport of 5-HT into the platelet (Lott et al. 1972; McCoy et al. 1974). However, the mechanism responsible for elevated blood 5-HT content in autism remains unknown. Platelet 5-HT uptake has been variably reported as normal (Boullin et al. 1971; Yuwiler et al. 1975) or increased (Katsui et al. 1986; Rotman et al. 1980) in autistic youngsters. The binding of imipramine to the 5-HT uptake site does not appear to differ in autistic and normal children (Anderson et al. 1984). Although an early report described increased 5-HT efflux from platelets of autistic children (Boullin et al. 1971), subsequent studies have failed to replicate this finding (Boullin et al. 1982; Yuwiler et al. 1975). Most studies have indicated that platelet number (Anderson et al. 1987; Geller et al. 1988; Yuwiler et al. 1975) and volume (Geller et al. 1988) are similar in autistic children and normal control subjects. Although a number of studies have shown that platelet MAO activity is not altered in autism (Campbell et al. 1976; Cohen et al. 1977b; Lake et al. 1977), 5-HT is primarily catabolized by MAO-A rather than MAO-B, the enzyme subtype found in platelets (Donnelly and Murphy 1977). Of note, whole-blood tryptophan levels appear to be normal in autism (Minderaa et al. 1987). Studies of peripheral 5-HT turnover in autistic subjects are considered in the next section of this chapter. Anderson et al. (1987), Hanley et al. (1977), and Yuwiler et al. (1985) provide further discussion of mechanisms that may potentially contribute to hyperserotonemia in autism.

### Serotonin Metabolites

*Cerebrospinal Fluid.* Measurement of cerebrospinal fluid (CSF) and urinary levels of the 5-HT metabolite 5-hydroxyindoleacetic acid (5-HIAA) has been used to assess the rate of 5-HT turnover in autism. CSF 5-HIAA concentrations appear to be normal or only slightly reduced in autism. In 1974, Cohen and colleagues reported that CSF levels of 5-HIAA following probenecid administration did not differ significantly between autistic children and children with other pervasive developmental disorders ("atypical children") or epilepsy, although more severely disturbed autistic children tended to have lower 5-HIAA levels than atypical children. In a subsequent study (Cohen et al. 1977a), the group found lower postprobenecid 5-HIAA concentrations in autistic youngsters than in nonautistic psychotic children, but did not detect a difference in levels of 5-HIAA or other CSF metabolites in autistic children vs. children with primary aphasia, youngsters with hyperactivity and poor attention span, or children who had undergone pediatric evaluation for a variety of neurological complaints. A third study, carried out by Winsberg et al. (1980), suggested that some autistic subjects may not show the expected increase in CSF 5-HIAA levels following probenecid administration; however, the absence of a control group limits the conclusions to be drawn. In studies that did not employ probenecid (Gillberg et al.

1983; Ross et al. 1985), baseline 5-HIAA concentrations did not differ in autistic children vs. age- and gender-matched pediatric control subjects with negative neurological workups or children with nonautistic psychotic disorders. No correlation was found between CSF 5-HIAA concentrations and IQ among either autistic or non-autistic psychotic subjects in the Gillberg study. To date, the relationship between blood 5-HT content and CSF 5-HIAA levels has not been evaluated in autistic subjects or normal control subjects.

*Urine.* Most 5-HIAA in urine, as in blood, is peripheral in origin. Three studies have shown similar baseline urinary 5-HIAA excretion in autistic subjects vs. normal control subjects (Minderaa et al. 1987; Schain and Freedman 1961; Shaw et al. 1959). Studies of urinary 5-HIAA levels following administration of the 5-HT precursor tryptophan have suggested lower (Sutton et al. 1958) or similar (Schain and Freedman 1961; Shaw et al. 1959) rates of 5-HIAA excretion in autistic and normal individuals. Although autistic individuals as a group do not appear to have elevated 5-HIAA excretion, evidence suggests that autistic subjects with hyperserotonemia may excrete somewhat more 5-HIAA than do normal individuals. Recently, Minderaa et al. (1987) reported significantly increased levels of 5-HIAA in the urine of four hyperserotonemic autistic subjects contrasted with 27 healthy control subjects. Hanley et al. (1977) found elevated urinary 5-HIAA excretion both before and after tryptophan loading in four hyperserotonemic autistic children who were compared with four normoserotonemic mildly retarded youngsters. While significant correlations have not been observed between whole-blood 5-HT content and urinary 5-HIAA excretion in autistic (Minderaa et al. 1987), severely retarded (Partington et al. 1973), or healthy subjects (Minderaa et al. 1987), a strong positive correlation between the two measures has been reported in a small sample comprised only of unmedicated hyperserotonemic autistic individuals (Minderaa et al. 1987). In summarizing the findings of the foregoing studies, Minderaa and colleagues (1987) concluded that increased synthesis of 5-HT in the bowel does not appear to be the primary cause of hyperserotonemia in autism, although a small increase in synthesis may be a contributing factor.

### Serotonin Receptor-Related Studies

*Antiserotonin Receptor Antibodies.* In 1985, Todd and Ciaranello described the isolation of antibodies to 5-HT$_{1A}$ receptors in human brain from both the blood and the CSF of one autistic child. The authors alluded to the presence of circulating antibodies to central

5-HT receptors in 7 of 13 children they had studied at the time. Anti–5-HT receptor antibodies were not detected in the blood of each of 13 normal control children. Although the report is of major interest because it suggests the possibility that an autoimmune process may impair central 5-HT receptor function in some autistic individuals, the finding of anti–5-HT receptor antibodies in autism has not been replicated to date.

*Platelet 5-HT2 Receptor Function.* 5-HT2 receptors have been demonstrated on the human blood platelet. The platelet receptor has drug affinities comparable to those of the 5-HT2 receptor in human frontal cortex (Geaney et al. 1984; McBride et al. 1983). Phosphoinositide hydrolysis appears to serve as the signal-transducing system for the platelet as well as the central 5-HT2 receptor (de-Chaffoy de Courcelles et al. 1985). Thus the platelet 5-HT2 receptor may serve as a peripheral model for the evaluation of 5-HT2 receptor function in disease states. The physiological responsivity of the platelet 5-HT2 receptor population can be assessed in conjunction with measurement of receptor binding indices, since 5-HT–induced and 5-HT–amplified platelet aggregation are mediated by the 5-HT2 receptor complex (DeClerck et al. 1982).

We have previously reported (McBride et al. 1989) the results of a pilot study of platelet 5-HT2 receptor binding indices and responsivity in seven drug-free male autistic young adults vs. eight age- and gender-matched normal control subjects. The autistic subjects were all verbal and relatively high functioning, with a mean full scale IQ of $74 \pm 6$ (SD). Platelet 5-HT2 receptor sites were labeled with $2[-^{125}\text{I}]$iodolysergic acid diethylamide ($[^{125}\text{I}]$ILSD), and specific binding defined with 1.0 μM ketanserin, a specific 5-HT2 receptor antagonist. 5-HT–amplified platelet aggregation responses were measured in the presence of adenosine diphosphate (ADP).

As shown in Figure 3-1, autistic subjects exhibited a lower mean number ($B_{max}$) of platelet 5-HT2 receptors than did healthy subjects. No significant difference in the dissociation constant ($K_D$) was observed between groups. The magnitude of 5-HT–amplified platelet aggregation responses, quantified by the "serotonin augmentation index" (SAI; McBride et al. 1987), was significantly decreased in autistic subjects vs. normal control subjects. In the case of most of the autistic subjects, decreased 5-HT–amplified aggregation responses might be explained by the paucity of platelet 5-HT2 receptors. However, alterations in the function of the receptor-linked signal-transducing system have not been ruled out.

As shown in Figure 3-2, the magnitude of the 5-HT–amplified

platelet aggregation response (quantified by the SAI) was negatively correlated with whole-blood 5-HT content in autistic subjects. The two autistic subjects with hyperserotonemia (whole-blood 5-HT content > 220 ng/ml) had particularly blunted platelet aggregation responses. It is unlikely that the negative correlation between the SAI and whole-blood 5-HT content reflects a modulatory response of the platelet 5-HT$^2$ receptor to exposure to higher concentrations of 5-HT in vivo. The receptor binding site is located on the external plasma membrane, whereas 5-HT in whole blood is compartmentalized in platelet cytoplasmic storage granules. Moreover, we and others have found that plasma free (G. M. Anderson, unpublished data; Cook et al. 1988) and urinary (Anderson et al. 1989) 5-HT levels do not differ in autistic vs. normal children. The data suggest the hypothesis that autistic subjects with elevated whole-blood 5-HT content may have more pronounced alterations in platelet 5-HT$^2$ receptor function than do normoserotonemic autistic subjects.

Additional studies are required not only to test the foregoing hypothesis, but to evaluate the specificity of altered platelet 5-HT$_2$ receptor function for autistic disorder vs. cognitive impairment. It will

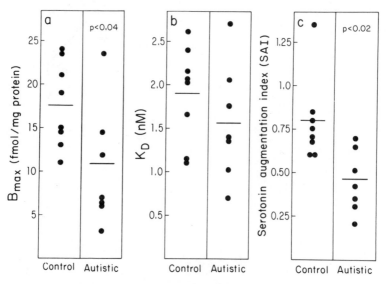

**Figure 3-1.** Values for platelet 5-HT$_2$ receptor indices in autistic ($n = 7$) vs. control subjects ($n = 8$), including (a) the mean number of receptors ($B_{max}$), (b) the dissociation constant ($K_D$), and (c) the serotonin augmentation index (SAI).

also be important to determine whether altered receptor function is present from an early age or develops later in life.

## Neuroendocrine Challenge Studies

Studies of metabolites in the CSF do not directly address the crucial question of whether neurotransmitter-mediated responses are altered

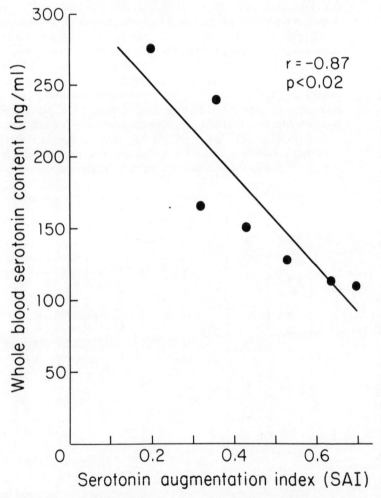

**Figure 3-2.** Correlation between the serotonin augmentation index (SAI) and whole-blood 5-HT content in autistic subjects ($N = 7$).

in neuropsychiatric disorders. Neuroendocrine challenge tests have been widely employed to assess the responsivity of serotonergic pathways in the CNS. Numerous studies in animals and humans have shown that plasma prolactin levels are increased following the administration of agents that augment central serotonergic transmission, including 5-HT precursors, 5-HT–releasing agents, and direct 5-HT receptor agonists (Meltzer et al. 1982; Mueller et al. 1986; Siever et al. 1984).

We recently reported the results of a pilot study in which central serotonergic responsivity was assessed in male autistic young adults by means of the fenfluramine challenge test (McBride et al. 1989). Administered as a single dose, fenfluramine serves as an indirect 5-HT agonist by stimulating release of 5-HT from storage granules in the presynaptic neuron and by blocking 5-HT reuptake (Borroni et al. 1983). Since both pre- and postsynaptic serotonergic processes contribute to the magnitude of fenfluramine-induced prolactin secretion, the fenfluramine challenge test provides a measure of the net responsivity of serotonin neuron circuits as a functional unit.

The fenfluramine challenge test was administered to the same seven drug-free, higher-functioning autistic men who participated in the studies of platelet 5-HT$_2$ receptor function described earlier in this chapter, and to seven of the eight normal control subjects. All subjects received both a 60-mg oral dose of fenfluramine and placebo administered under double-blind conditions. Blood samples for measurement of plasma prolactin levels were drawn immediately preceding ingestion of fenfluramine or placebo, and hourly for five successive hours.

Fenfluramine ingestion resulted in a significant increase in plasma prolactin levels in both the autistic group and the control group. A repeated-measures analysis of variance (ANOVA) comparing fenfluramine-induced changes in plasma prolactin levels in autistic vs. control subjects indicated significantly decreased prolactin release in the autistic group ($F = 8.14$, $P < .02$). Figure 3-3a depicts the magnitude of individual prolactin responses to fenfluramine challenge, as quantified by the calculation of the area under the curve (AUC$_{\Delta PRL}$) using the trapezoid rule. The mean value for AUC$_{\Delta PRL}$ in the autistic group was less than half that in the control group.

Summed plasma levels of fenfluramine and its active metabolite norfenfluramine were significantly higher in autistic subjects than in normal control subjects. Thus poor absorption of fenfluramine was not the cause of decreased prolactin release in autistic individuals. When prolactin responses were adjusted for a subject's plasma levels of fenfluramine and norfenfluramine, differences between autistic and control subjects were enhanced. As shown in Figure 3-3b, all autistic subjects had lesser drug-adjusted prolactin responses than the least responsive control subject. The one autistic subject who exhibited an unadjusted response well within the normal range had plasma drug

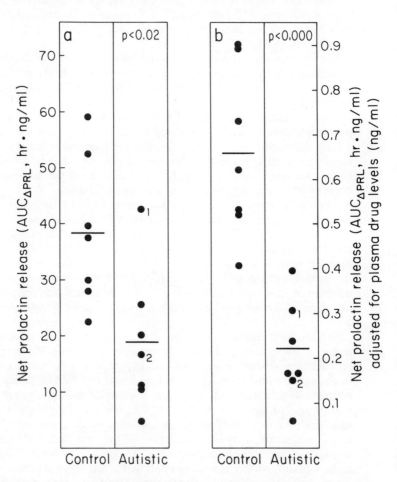

**Figure 3-3.** (a) Net fenfluramine-induced prolactin release, quantified by the area under the curve, in autistic ($n = 7$) versus control subjects ($n = 7$). (b) Net fenfluramine-induced prolactin release following adjustment for plasma fenfluramine and nor-fenfluramine levels. Subscripts "1" and "2" indicate subjects with unusually high plasma drug levels.

levels 1.9 times greater than the highest levels found in a control subject.

A repeated-measures ANOVA indicated that mean plasma prolactin levels during placebo trials were not significantly different in the autistic and control groups. Significant correlations were not found between baseline prolactin levels prior to fenfluramine administration and drug-induced prolactin responses.

To our knowledge, only one other published report has described neuroendocrine responses to serotonergic challenge agents in autistic subjects. Hoshino and colleagues (1984) observed blunted prolactin release following administration of the 5-HT precursor L-5-hydroxytryptophan (5-HTP) to six prepubertal autistic children vs. six hyperactive children and nine healthy young adults. Although the presence of an age-matched healthy control group would have enhanced these authors' study, the results are qualitatively similar to those of our study.

The results of the above neuroendocrine challenge studies are consistent with the hypothesis that central serotonergic responsivity is decreased in at least some individuals with autistic disorder. Blunted prolactin responses in autistic subjects might reflect decreased stores of 5-HT available for release from the presynaptic neuron, decreased numbers of postsynaptic $5-HT_1$ and/or $5-HT_2$ receptors, or impaired coupling of receptor-linked signal-transducing systems.

While neuroendocrine challenge studies have provided further data in support of altered serotonergic function in autism, findings must be interpreted cautiously. Serotonergic function in the hypothalamus may not necessarily parallel that in brain regions most involved in the pathogenesis of autism. Furthermore, the basal activity of other neurotransmitter systems may affect the magnitude of 5-HT–mediated prolactin release. Particular consideration must be devoted to the possibility that blunted prolactin responses in autistic subjects reflect increased tuberoinfundibular dopaminergic activity. However, existing neurochemical studies have generally not provided evidence of altered dopaminergic function in autism (Anderson 1987; Yuwiler et al. 1985). Although changes in the functional status of the lactotroph have not been excluded, a report (Suwa et al. 1984) of normal to enhanced prolactin secretion in response to thyrotropin-releasing hormone in four autistic youngsters suggests that the secretory capacity of the lactotroph is not impaired in autistic children.

## Correlations Between Measures of Central and Peripheral Serotonergic Responsivity in Autism

In our forementioned study (McBride et al. 1989), the magnitude of the 5-HT–amplified platelet aggregation response (quantified by the SAI) correlated positively with fenfluramine-induced prolactin release in autistic subjects (see Figure 3-4a). Furthermore, among autistic subjects, whole-blood 5-HT content correlated negatively with net prolactin release when values for $AUC_{\Delta PRL}$ were adjusted for plasma drug levels (see Figure 3-4b).

While the preliminary status of these correlations between measures of central and peripheral serotonergic function must be stressed, the correlations are of interest because they suggest the possibility of systemic alterations in serotonergic function in autistic disorder. The positive correlation between fenfluramine-induced prolactin release and the magnitude of the 5-HT–amplified platelet aggregation response raises the question of whether a decrease in the number or responsivity of hypothalamic 5-HT2 receptor complexes may play a role in mediating blunted prolactin responses in autistic subjects. In addition, the negative correlation between whole-blood 5-HT content and the drug-adjusted prolactin response to fenfluramine is consistent with the hypothesis that common processes may contribute

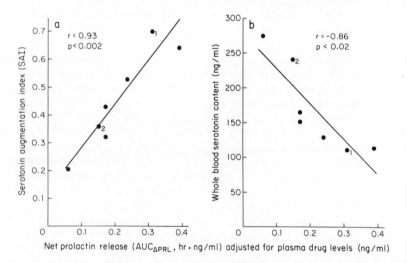

**Figure 3-4.**  Correlations in autistic subjects ($N = 7$) between net fenfluramine-induced prolactin release, following adjustment for plasma fenfluramine and norfenfluramine levels, and (a) the serotonin augmentation index (SAI); and (b) whole-blood 5-HT content. Subscripts "1" and "2" indicate subjects with unusually high plasma drug levels.

to the regulation of blood 5-HT levels and central serotonergic responsivity in autism. Our capacity to develop a more specific hypothesis is limited by the fact that the mechanism responsible for higher mean whole-blood 5-HT concentrations in autistic populations has not been established.

## TREATMENT STUDIES EMPLOYING SEROTONERGIC AGENTS

### Fenfluramine

Several studies (August et al. 1985; Campbell et al. 1986; Klykylo et al. 1985; Ritvo et al. 1983, 1986; Stubbs et al. 1986) have reported substantial clinical improvement in at least one-quarter of autistic children treated for several weeks with fenfluramine, a drug that depletes 5-HT stores in both the CNS (Reuter 1975) and blood (Ritvo et al. 1983) with chronic administration. Most of these studies have described gains in social relatedness and attention span and decreased hyperactivity and motor stereotypies. While some reports have described enhanced intellectual functioning following fenfluramine treatment (Geller et al. 1982; Ritvo et al. 1983), others have indicated a failure to replicate this finding (August et al. 1985; Klykylo et al. 1985).

The use of fenfluramine as a treatment for autism remains controversial. Data from a multicenter study indicate that a majority of autistic youngsters do not have a "strong" positive response to the drug (Ritvo et al. 1986). Furthermore, several authors have voiced concerns about side effects or a risk of neurotoxicity from long-term administration (Gualtieri 1986; Schuster et al. 1986; Volkmar et al. 1983b).

The apparent positive clinical response of some autistic children to fenfluramine treatment does not provide information about central serotonergic activity in such individuals. Animal studies suggest that the drug can stimulate serotonergically mediated prolactin secretion despite a substantial reduction in brain 5-HT content (Willoughby et al. 1982). Children with normal blood 5-HT levels actually seem to respond to fenfluramine better than children with initially elevated levels (Ritvo et al. 1986).

### Other Drugs

Campbell et al. (1987) have provided a comprehensive review of pharmacological treatment studies in autism. Most studies undertaken to date have evaluated the efficacy of antipsychotic agents or fenfluramine. The finding of blunted serotonergically mediated responses in autistic young adults (McBride et al. 1989) raises the question of whether drugs that augment central serotonergic transmission would be beneficial in the treatment of autistic individuals. Unfortunately, data are not currently available to address this question. Existing studies of the efficacy of imipramine (Campbell et al. 1971) and 5-HTP (Sverd et al. 1978) are inconclusive because of methodological limitations, including diagnostic imprecision, very small sample size, and the absence of plasma drug levels. Of interest, Campbell and colleagues reported "marked improvement" in 3 of 10 preschool-age "autistic and schizophrenic children" treated with 12.5 to 75 mg of imipramine per day for several weeks. However, seven additional children who received similar doses were only "slightly improved" or "worse" at the conclusion of the trial. Each of five children described as worse had shown initial improvement lasting for up to 2 weeks. Side effects limited the dose of 5-HTP administered to the three autistic children in the Sverd study, as well as the duration of treatment.

## SUMMARY

Evidence suggests that alterations in serotonergic function are associated with autistic disorder. The presence of higher mean levels of 5-HT in whole blood is the most firmly established neurochemical finding to date in autistic children. Although CSF levels of the 5-HT metabolite 5-HIAA appear to be essentially normal in autism, the results of existing neuroendocrine challenge studies are consistent with the hypothesis that central serotonergic responsivity is reduced. Furthermore, a recent study (McBride et al. 1989) reported that the magnitude of 5-HT–amplified platelet aggregation, a physiological response mediated by the platelet $5-HT_2$ receptor complex, is decreased in autistic subjects. As technologies become available, studies should be undertaken to evaluate the functional status of central 5-HT receptors in autism.

A report of significant correlations among whole-blood 5-HT content, fenfluramine-induced prolactin release, and the magnitude of 5-HT–amplified platelet aggregation responses in seven autistic young adults suggests a systemic alteration in serotonergic function in autism. However, this possibility has yet to be established. Additional studies in which multiple serotonergic markers are simultaneously assessed in the same subjects are needed. Research efforts must also be directed toward elucidation of the mechanisms responsible for altered serotonergic measures in the CNS and periphery.

Finally, neurotransmitter systems other than 5-HT may play a role in the pathogenesis of autism, although studies to date have not definitively shown changes in either dopaminergic or noradrenergic function (Anderson 1987; Yuwiler et al. 1985). Alterations in the function of one neurotransmitter may impact upon the function of other systems. For instance, Cohen and Young (1977) have hypothesized that underactivity of inhibitory central serotonergic pathways may result in functional dopaminergic overactivity. Studies that simultaneously assess the status of more than one neurotransmitter may help to identify imbalances in the relative activities of neurotransmitter systems.

# REFERENCES

American Psychiatric Association: Diagnostic and Statistical Manual of Mental Disorders, 3rd Edition. Washington, DC, American Psychiatric Association, 1980

American Psychiatric Association: Diagnostic and Statistical Manual of Mental Disorders, 3rd Edition, Revised. Washington, DC, American Psychiatric Association, 1987

Anderson GM: Monamines in autism: an update of neurochemical research on a pervasive developmental disorder. Med Biol 65:67–74, 1987

Anderson GM, Minderaa RB, van Bentem PPG, et al: Platelet imipramine binding in autistic subjects. Psychiatry Res 11:133–141, 1984

Anderson GM, Freedman DX, Cohen DJ, et al: Whole blood serotonin in autistic and normal subjects. J Child Psychol Psychiatry 28:885–900, 1987

Anderson GM, Minderaa RB, Cho SC, et al: The issue of hyperserotonemia and platelet serotonin exposure: a preliminary study. J Autism Dev Disord 19:349–351, 1989

August GJ, Raz N, Baird TD: Brief report: effects of fenfluramine on behavioral, cognitive, and affective disturbances in autistic children. J Autism Dev Disord 15:97–107, 1985

Borroni E, Ceci A, Garattini S: Differences between d-fenfluramine and d-norfenfluramine in serotonin presynaptic mechanisms. J Neurochem 40:891–893, 1983

Boullin DJ, Coleman M, O'Brien RA, et al: Laboratory prediction of infantile autism based on 5-hydroxytryptamine efflux from blood platelets and their correlation with the Rimland E-2 Score. J Autism Childhood Schizophr 1:63–71, 1971

Boullin DJ, Freeman BJ, Geller E, et al: Toward the resolution of conflicting findings. J Autism Dev Disord 12:97–98, 1982

Brown WT, Jenkins EC, Friedman E, et al: Autism is associated with the fragile-X syndrome. J Autism Dev Disord 12:303–308, 1982

Campbell M, Fish B, Shapiro T, et al: Imipramine in preschool autistic and schizophrenic children. J Autism Childhood Schizophr 1:267–282, 1971

Campbell M, Friedman E, Green WH, et al: Blood serotonin in schizophrenic children. Int Pharmacopsychiatry 10:213–221, 1975

Campbell M, Friedman E, Green WH, et al: Blood platelet monoamine oxidase activity in schizophrenic children and their families. Neuropsychobiology 2:239–246, 1976

Campbell M, Perry R, Polonsky BR, et al: Brief report: an open report of fenfluramine in hospitalized young autistic children. J Autism Dev Disord 16:495–506, 1986

Campbell M, Anderson LT, Green WH, et al: Psychopharmacology, in Handbook of Autism and Pervasive Developmental Disorders. Edited by Cohen DJ, Donnellan AM, Paul R. New York, John Wiley, 1987, pp 545–565

Cohen DJ, Young JG: Neurochemistry and child psychiatry. J Am Acad Child Psychiatry 16:353–411, 1977

Cohen DJ, Shaywitz BA, Johnson WT, et al: Biogenic amines in autistic and atypical children. Arch Gen Psychiatry 31:845–853, 1974

Cohen DJ, Caparulo BK, Shaywitz BA, et al: Dopamine and serotonin metabolism in neuropsychiatrically disturbed children. Arch Gen Psychiatry 34:545–550, 1977a

Cohen DJ, Young JG, Roth JA: Platelet monoamine oxidase in early childhood autism. Arch Gen Psychiatry 34:534–537, 1977b

Cook EH, Leventhal BL, Freedman DX: Free serotonin in plasma: autistic children and their first-degree relatives. Biol Psychiatry 24:488–491, 1988

Courchesne E, Yeung-Courchesne R, Press GA, et al: Hypoplasia of cerebellar vermal lobules VI and VII in autism. N Engl J Med 318:1349–1354, 1988

Crawford N, Sutton M, Horsfield GI: Platelets in the carcinoid syndrome: a chemical and ultrastructural investigation. Br J Haematol 13:181–188, 1967

deChaffoy de Courcelles D, Leysen JE, DeClerck F, et al: Evidence that phospholipid turnover is the signal transducing system coupled to serotonin-S2 receptor sites. J Biol Chem 260:7603–7608, 1985

DeClerck F, David JL, Janssen PAJ: Inhibition of 5-hydroxytryptamine-induced and -amplified human platelet aggregation by ketanserin (R41468), a selective 5-HT2-receptor antagonist. Agents Actions 12:388–397, 1982

DeLisi LE, Neckers LM, Weinberger DR, et al: Increased whole blood serotonin concentrations in chronic schizophrenic patients. Arch Gen Psychiatry 38:647–650, 1981

Donnelly CH, Murphy DL: Substrate- and inhibitor-related characteristics of human platelet monoamine oxidase. Biochem Pharmacol 26:853–858, 1977

Geaney DP, Schachter M, Elliot JM, et al: Characterization of [$^3$H]lysergic acid diethylamide binding to a 5-hydroxytryptamine receptor on human platelet membranes. Eur J Pharmacol 97:87–93, 1984

Geller E, Ritvo ER, Freeman BJ, et al: Preliminary observations on the effect of fenfluramine on blood serotonin and symptoms in three autistic boys. N Engl J Med 307:165–169, 1982

Geller E, Yuwiler A, Freeman BJ, et al: Platelet size, number, and serotonin content in blood of autistic, childhood schizophrenic, and normal children. J Autism Dev Disord 18:119–127, 1988

Gillberg C, Svennerholm L, Hamilton-Hellberg C: Childhood psychosis and monoamine metabolites in spinal fluid. J Autism Dev Disord 13:383–396, 1983

Gualtieri CT: Fenfluramine and autism: careful reappraisal is in order. J Pediatr 108:417–419, 1986

Hanley HG, Stahl SM, Freedman DX: Hyperserotonemia and amine metabolism in autistic and retarded children. Arch Gen Psychiatry 34:521–531, 1977

Hoshino Y, Tachibana R, Watanabe M, et al: Serotonin metabolism and hypothalamic-pituitary function in children with infantile autism and minimal brain dysfunction. Jpn J Psychiatry Neurol 26:937–945, 1984

Katsui T, Okuda M, Usuda S, et al: Kinetics of $^3$H-serotonin uptake by platelets in infantile autism and developmental language disorder (including five pairs of twins). J Autism Dev Disord 16:69–76, 1986

Klykylo WM, Feldis D, O'Grady D, et al: Brief report: clinical effects of fenfluramine in ten autistic subjects. J Autism Dev Disord 15:417–423, 1985

Lake CR, Ziegler MG, Murphy DL: Increased norepinephrine levels and decreased DBH in primary autism. Arch Gen Psychiatry 35:553–556, 1977

Lott IT, Chase TN, Murphy DL: Down's syndrome: transport, storage, and metabolism of serotonin in blood platelets. Pediatr Res 6:730–735, 1972

Mann JJ, Stanley M, McBride PA, et al: Increased serotonin and β-adrenergic receptor binding in the frontal cortices of suicide victims. Arch Gen Psychiatry 43:954–959, 1986

McBride PA, Mann JJ, McEwen B, et al: Characterization of serotonin binding sites on human platelets. Life Sci 33:2033–2041, 1983

McBride PA, Mann JJ, Polley MJ, et al: Assessment of binding indices and physiological responsiveness of the 5-HT$_2$ receptor on human platelets. Life Sci 40:1799–1809, 1987

McBride PA, Anderson GM, Hertzig ME, et al: Serotonergic responsivity in male young adults with autistic disorder: results of a pilot study. Arch Gen Psychiatry 46:213–221, 1989

McCoy EE, Segal DJ, Bayer SM, et al: Decreased ATPase and increased sodium content of platelets of Down's syndrome: relation to decreased serotonin content. N Engl J Med 291:950–953, 1974

Meltzer HY, Wiita B, Tricou BJ, et al: Effect of serotonin precursors and serotonin agonists on plasma hormone levels, in Serotonin in Biological Psychiatry. Edited by Ho BT, Schoolar JC, Usdin E. New York, Raven Press, 1982, pp 117–139

Minderaa RB, Anderson GM, Volkmar FR, et al: Urinary 5-hydroxyindoleacetic acid and whole blood serotonin and tryptophan in autistic and normal subjects. Biol Psychiatry 22:933–940, 1987

Morgan DG, May PC, Finch CE: Dopamine and serotonin systems in human and rodent brain: effects of age and neurodegenerative disease. J Am Geriatr Soc 35:334–345, 1987

Mueller EA, Murphy DL, Sunderland T: Further studies of the putative serotonin agonist, m-chlorophenylpiperazine: evidence for a serotonin receptor mediated mechanism of action in humans. Psychopharmacology (Berlin) 89:388–391, 1986

Ornitz EM, Ritvo ER: The syndrome of autism: a critical review. Am J Psychiatry 133:609–621, 1976

Partington MW, Tu JB, Wong CY: Blood serotonin levels in severe mental retardation. Dev Med Child Neurol 15:616–627, 1973

Prior M, Tress B, Hoffman W, et al: Computed tomographic study of children with classic autism. Arch Neurol 41:482–484, 1984

Reuter CJ: A review of the CNS effects of fenfluramine, 780SE and norfenfluramine on animals and man. Postgrad Med 51 (suppl 1):18–27, 1975

Ritvo ER, Yuwiler A, Geller E, et al: Increased blood serotonin and platelets in early infantile autism. Arch Gen Psychiatry 23:566–572, 1970

Ritvo ER, Freeman BJ, Geller E, et al: Effects of fenfluramine on 14 outpatients with the syndrome of autism. J Am Acad Child Psychiatry 22:549–558, 1983

Ritvo ER, Freeman BJ, Yuwiler A, et al: Fenfluramine treatment of autism: UCLA collaborative study of 81 patients at nine medical centers. Psychopharmacol Bull 22:133–140, 1986

Rosenbloom S, Campbell M, George AE, et al: High resolution CT scanning in infantile autism: a quantitative approach. J Am Acad Child Psychiatry 23:72–77, 1984

Ross DL, Klykylo WM, Anderson GM: Cerebrospinal fluid indoleamine and monoamine effects of fenfluramine treatment of infantile autism. Ann Neurol 18:394, 1985

Rotman A, Caplan R, Szekeley GA: Platelet uptake of serotonin in psychotic children. Psychopharmacology (Berlin) 67:245–248, 1980

Rutter M: Infantile autism, in The Clinical Guide to Child Psychiatry. Edited by Shaffer D, Ehrhardt A, Greenhill L. New York, Free Press, 1985, pp 48–78

Schain RJ, Freedman DX: Studies on 5-hydroxyindole metabolism in autistic and other mentally retarded children. J Pediatr 58:315–320, 1961

Schuster CR, Lewis M, Seiden LS: Fenfluramine: neurotoxicity. Psychopharmacol Bull 22:148–151, 1986

Shaw CR, Lucas J, Rabinovitch RD: Metabolic studies in childhood schizophrenia. Arch Gen Psychiatry 1:366–371, 1959

Siever LJ, Murphy DL, Slater S, et al: Plasma prolactin changes following fenfluramine in depressed patients compared to controls: an evaluation of central serotonergic responsivity in depression. Life Sci 34:1029–1039, 1984

Siever LJ, Coccaro EF, Klar H, et al: Aminergic measures in personality disorders, in Proceedings of the IVth World Congress of Biological Psychiatry. Edited by Shagass C. New York, Elsevier, 1986, pp 264–266

Sneddon JM: Blood platelets as a model for monoamine-containing neurones. Prog Neurobiol 1:151–198, 1973

Stubbs EG, Budden SS, Jackson RH, et al: Effects of fenfluramine on eight outpatients with the syndrome of autism. Dev Med Child Neurol 28:229–235, 1986

Sutton HE, Read JH, Arbor A: Abnormal amino acid metabolism in a case suggesting autism. Am J Dis Child 96:23–28, 1958

Suwa S, Naruse H, Ohura T, et al: Influence of pimozide on hypothalamo-pituitary function in children with behavioral disorders. Psychoneuroendocrinology 9:37–44, 1984

Sverd J, Kupietz SS, Winsberg BG: Effects of L-5-hydroxytryptophan in autistic children. J Autism Child Schizophr 8:171–180, 1978

Takahashi S, Kanai H, Miyamoto Y: Reassessment of elevated serotonin levels in blood platelets in early infantile autism. J Autism Child Schizophr 6:317–326, 1976

Thomas DP, Vane JR: 5-Hydroxytryptamine in the circulation of the dog. Nature 216:335–338, 1967

Todd RD, Ciaranello RD: Demonstration of inter- and intraspecies differences in serotonin binding sites by antibodies from an autistic child. Proc Natl Acad Sci USA 82:612–616, 1985

Tu J, Partington MW: 5-Hydroxyindole levels in the blood and CSF in Down's syndrome, phenylketonuria and severe mental retardation. Dev Med Child Neurol 14:457–466, 1972

van Praag HM: Indoleamines in depression and suicide. Prog Brain Res 65:59–71, 1986

Volkmar FR, Anderson GM, Hoder EL, et al: The relation of serotonin level to behavioral characteristics in autistic children. Paper presented at the annual meeting of the American Academy of Child Psychiatry, San Francisco, CA, October 1983a

Volkmar FR, Paul R, Cohen DJ, et al: Irritability in autistic children treated with fenfluramine. N Engl J Med 309:187, 1983b

Willoughby JO, Menadue M, Jervois P: Function of serotonin in the physiologic secretion of growth hormone and prolactin; action of 5,7-dihydroxytryptamine, fenfluramine and p-chlorophenylalanine. Brain Res 249:291–299, 1982

Winsberg BG, Sverd J, Castells S, et al: Estimation of monoamine and cyclic-AMP turnover and amino acid concentrations of spinal fluid in autistic children. Neuropediatrics 11:250–255, 1980

Yuwiler A, Ritvo E, Geller E, et al: Uptake and efflux of serotonin from platelets of autistic and nonautistic children. J Autism Childhood Schizophr 5:83–98, 1975

Yuwiler A, Geller E, Ritvo E: Biochemical studies of autism, in Handbook of Neurochemistry. Edited by Lajtha E. New York, Plenum, 1985, pp 671–691

# Chapter 4

## *Serotonin in Mood and Personality Disorders*

Emil F. Coccaro, M.D.
Larry J. Siever, M.D.
Kim R. Owen, M.D.
Kenneth L. Davis, M.D.

# Chapter 4

# *Serotonin in Mood and Personality Disorders*

N eurochemical and neuropharmacologic evidence from over two decades of research suggests that dysfunction of central serotonergic systems plays a significant role in disorders of mood and personality. Because serotonin (5-hydroxytryptamine [5-HT]) generally appears to exert an inhibitory effect on neuronal firing, central serotonergic systems are thought to be integrally involved in the processing of various incoming stimuli. Accordingly, behavioral and neurophysiological evidence of the role of 5-HT in the homeostatic maintenance of arousal, mood, appetite, sleep, and other neurobiological functions (Voyt 1982), is consistent with clinical observations suggesting a central disturbance of these functions in patients with disorders of mood and personality.

This chapter will review clinical psychobiologic data relevant to the role of the central serotonergic system in patients with major mood and/or personality disorder. Additional discussions regarding the role of central 5-HT in modulating certain specific behaviors (e.g., suicide, violence, impulsiveness), and during treatment with pharmacologic agents that may affect the central serotonergic system, may be found elsewhere in this volume (see Chapters 8 and 10).

## SEROTONERGIC STUDIES OF MOOD DISORDER

**Central Indices of Serotonin**

*Lumbar Cerebrospinal Fluid Serotonin Metabolite Concentration.* Despite the potential difficulties in interpreting data from studies of lumbar cerebrospinal fluid (CSF) concentrations of the 5-HT metabolite 5-hydroxyindoleacetic acid (5-HIAA) in man (Kuhn et al. 1986), many studies support the hypothesis that presynaptic central 5-HT activity, as reflected by CSF 5-HIAA concentration, is diminished in a subgroup of patients with major mood disorders. Reduced CSF 5-HIAA concentrations have been reported in several studies of drug-free depressed patients (Agren 1980; Åsberg et al. 1976a, 1984). While other studies have failed to replicate these findings, the sample sizes employed in these studies were not large enough to detect even moderately sized (i.e., $\geq 0.50$ $z$ score) reductions in CSF 5-HIAA, and on average had less than a 30% chance of detecting such reductions in this metabolite (i.e., mean of $26 \pm 14\%$ in the 10 negative studies examined in Rothpearl et al. 1981). Alternatively, the failure to identify reduced CSF 5-HIAA concentrations in depressed patients may be due to the presence of a central serotonergic abnormality in only a subgroup of depressed patients. A subgroup of depressed patients with "low" CSF 5-HIAA concentrations was first reported by Åsberg et al. (1976a). This "low" 5-HIAA group differed from the "normal" 5-HIAA group by the finding of an inverse relationship between CSF levels of the metabolite and severity of depression, a relationship that was not present in the "normal" 5-HIAA group. The presence of a subgroup of depressed patients with "low" CSF 5-HIAA was confirmed in a reanalysis of data from two large independent samples of depressed patients and nondepressed control subjects (Gibbons and Davis 1986) in which CSF 5-HIAA values were resolved into a bimodal distribution. In this distribution, approximately 35% of depressed patients in either sample fell into a "low" CSF 5-HIAA mode of nearly equal value in both samples (means of natural log [ln] 4.68 ng/ml and ln 4.63 ng/ml for the two depressed patient samples, respectively).

Behavioral correlates of reduced lumbar CSF 5-HIAA concentrations in depressed patients are inconsistent. Inverse relationships between CSF 5-HIAA concentrations and severity of depression (Åsberg et al. 1976a) and/or specific depressive symptoms (Agren 1980; Roy et al. 1985) have been reported in some but not all studies (Davis et al. 1981). More consistent, in this regard, is an inverse association between CSF 5-HIAA concentrations and history of suicide attempt in unipolar (but not bipolar) depressed patients (Åsberg et al. 1987; also see Chapter 8). Reduced lumbar CSF 5-HIAA concentrations in suicide-attempting patients with diagnoses other than depression have also been reported, suggesting that this behavioral correlate may not be specific to depression (Åsberg et al. 1987). Accordingly, it is possible that the reduced lumbar CSF 5-HIAA concentrations associated with history of suicide attempt may themselves account for a significant proportion of the depressed patients falling into the "low" CSF 5-HIAA mode (Meltzer and Lowy 1987).

Cerebrospinal fluid 5-HIAA may represent a state-independent characteristic in volunteer subjects and psychiatric patients. Stable concentrations of CSF 5-HIAA in healthy volunteers and in depressed patients examined between two episodes of depression have been reported (Träskman-Bendz et al. 1984). CSF 5-HIAA concentrations may increase modestly from depression to recovery in some depressed patients, but only if the concentrations were low when these patients were acutely ill (Träskman-Bendz et al. 1984). This increase in CSF 5-HIAA concentrations with recovery was only modest in magnitude, however, suggesting that dysregulation of central 5-HT may persist into the well state. Instability of central 5-HT function, as manifested by a persistent reduction in CSF 5-HIAA concentration (post-probenecid) across depressive and euthymic states, was associated with higher risk for subsequent depression in one study (van Praag and de Haan 1979). Moreover, subsequent treatment with the 5-HT precursor 5-hydroxytryptophan (5-HTP) (as compared with placebo) appeared to prevent the occurrence of another depressive episode in these patients (van Praag and de Haan 1980). Thus, reduced central 5-HT activity may represent a vulnerability factor for future depressive episodes in some patients with depressive disorder.

***Serotonin Pharmacochallenge Studies.*** Studies of neuroendocrine responses to agents that act to increase central 5-HT activity offer the potential advantage of examining the physiologic responsivity of synaptic 5-HT receptors as well as of the system as a whole. While differences in measures of presynaptic (i.e., decreased CSF 5-HIAA concentrations, brain tritiated-imipramine binding) and postsynaptic (i.e., increased brain 5-HT$_2$ receptor binding; see below) indices suggest diminished central 5-HT activity in depressed patients, the actual "net" activity of the system cannot be assessed by such indices alone. Reduced lumbar CSF 5-HIAA concentrations may indicate decreased 5-HT neurotransmission, but only if postsynaptic receptor sensitivity does not sufficiently increase to compensate for the reduced presynaptic activity. While increased 5-HT$_2$ receptor binding in the brains of suicide victims has been reported (Aurora and Meltzer 1989; Ferrier et al. 1986; Mann et al. 1986; Stanley and Mann 1983), ligand binding measures cannot reflect the physiologic responsivity of central 5-HT receptors and cannot be used to determine whether a compensation for the presynaptic reductions in 5-HT activity has in fact occurred.

Neuroendocrine studies of depressed patients have focused on the prolactin and cortisol responses to challenge with central serotonergic agents. Acute challenge with a variety of central serotonergic agents releases prolactin and/or ACTH/cortisol in humans in a dose-responsive fashion (see Chapter 1). Release of these hormones is thought to be mediated by stimulation of 5-HT receptors on hypothalamic cells, which release hypothalamic releasing factors (e.g., prolactin-releasing factor, corticotropin-releasing factor) that in turn stimulate specific anterior pituitary cells to release prolactin and/or ACTH (which in turn stimulates the release of cortisol from the adrenal cortex).

Both intravenously administered tryptophan (a 5-HT precursor) and orally administered fenfluramine (a 5-HT releaser/uptake inhibitor) stimulate prolactin and have been utilized in investigations of mood disorder patients. While the action of both agents involve presynaptic mechanisms, prolactin responses to both may involve postsynaptic mechanisms, particularly in the case of fenfluramine, whose metabolite may have direct 5-HT receptor agonist activity (Invernizzi et al. 1982). These agents thus have the advantage of simultaneously assessing pre- and postsynaptic (i.e., "net") aspects of central 5-HT activity in the hypothalamo-pituitary axis. Despite this similarity, fenfluramine challenge may be the more physiologic of the two paradigms because it releases endogenous stores of 5-HT rather than newly synthesized 5-HT.

Reduced prolactin responses to these central serotonergic probes have been found in many but not all studies. Prolactin responses to tryptophan infusion are reduced in drug-free depressed patients compared to nondepressed control subjects in all studies published to date (Cowen and Charig 1987; Heninger et al. 1984; Koyama and Meltzer 1986). However, the reduced prolactin response to tryptophan may in part be due to altered disposition of tryptophan leading to reduced postinfusion blood levels of tryptophan in depressed patients (Koyama and Meltzer 1986). Conversely, significant weight loss (frequently observed in depressed patients) has been shown to artifactually elevate prolactin responses to tryptophan infusion (Cowen and Charig 1987), suggesting that the prolactin response to this agent may underestimate the proportion of patients with reduced central 5-HT function.

Reduced prolactin responses to fenfluramine were found in patients with major depression compared to healthy control subjects in one study (Siever et al. 1984). However, a recent study of depressed outpatients and nondepressed control subjects did not replicate this finding (Asnis et al. 1988). Among other explanations, Asnis et al. (1988) suggested that reduced prolactin responses to fenfluramine may be confined to depressed patients with moderately severe illness. This possibility is supported by a third study, which reported reduced prolactin responses to fenfluramine in patients with uncomplicated major depression (e.g., without concomitant history of panic anxiety) compared to depressed patients with panic attacks or to patients with dysthymic disorder (Lopez-Ibor et al. 1988).

Reduced prolactin responses to central serotonergic probes do not appear to be related to severity of the depressive state, however, because prolactin responses to these agents have not been shown to correlate with depression rating scores in depressed patients. Thus, reduced prolactin responses to these agents may represent a relatively state-independent correlate of mood disorder.

Studies from our laboratory that were designed to address this possibility replicated earlier findings of reduced prolactin responses to fenfluramine in drug-free acutely depressed patients compared with drug-free nondepressed control subjects. Drug-free depressed patients who were in remission (including a subsample studied drug-free in both the acute and the remitted states) displayed reduced prolactin responses to fenfluramine similar to the reduction in responses seen in the acutely depressed patients (Figure 4-1). While reduced prolactin responses to fenfluramine were not characteristic of all patients, the proportion of subjects with significantly reduced, or "blunted," prolactin responses to fenfluramine (i.e., $\leq 7$ ng/ml) was significantly greater among depressed patients regardless of state (10 of 25, 40% vs. 1 of 18, 6% among nondepressed control subjects; $P < .05$) and regardless of a past history of suicide attempt (6 of 18, 33% vs. 6% among nonsuicide-attempting healthy control subjects; $P < .05$). This proportion is similar to that reported for depressed patients demonstrating reduced CSF concentrations of 5-HIAA (Gibbons and Davis 1986) and supports the hypothesis that diminished central 5-HT function may be characteristic of only a subgroup of patients with a major mood disorder.

Definitive statements regarding the state-independence of reduced neuroendocrine responses to central serotonergic probes await the results of studies examining a large number of patients across mood states. However, preliminary data from our studies suggest that prolactin responses to fenfluramine, like measures of CSF 5-HIAA, may be relatively characteristic of drug-free depressed patients across mood states. Examination of six patients studied drug-free in both the acute and the remitted states revealed a high correlation between prolactin responses to fenfluramine across states ($r = .93$, $P < .01$). While significant increases in prolactin responses to fenfluramine were noted in these patients as a group ($P < .05$), the percent changes across states were highly variable (range: 0 to 68%; mean: $42 \pm 27\%$), with two of the six patients displaying little or no change across states (Figure 4-1). Altered cortisol responses to 5-HTP challenge (see below) have also been reported in both depressed and manic patients

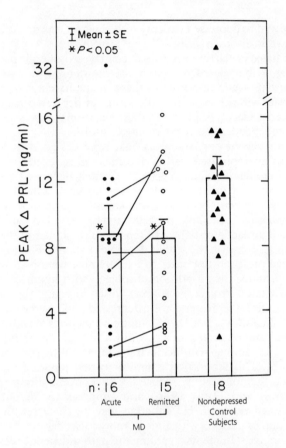

**Figure 4-1.**  Peak change in prolactin (peak ΔPRL [ng/ml]) responses to fenfluramine challenge in 16 male depressed patients studied in the drug-free acute state (closed circles; aged 55.4 ± 9.4 years), in 15 male depressed patients studied in the drug-free remitted state (open circles; aged 54.9 ± 7.7 years), and in 18 male drug-free nondepressed control volunteers (closed triangles; aged 45.9 ± 14.6 years). Reduced prolactin responses to fenfluramine ($P < .05$) were also noted in a subgroup of 10 acute ($n = 6$) and remitted ($n = 4$) depressed patients (6.5 ± 3.7 ng/ml; aged 53.7 ± 10.6 years; weight: 77.3 ± 14.9 kg) who were matched closely for age and weight with 10 nondepressed control subjects (12.4 ± 8.0 ng/ml; aged 52.3 ± 12.3 years; weight: 77.0 ± 12.7 kg). Six depressed patients studied in two drug-free states are indicated by the connecting lines.

(Meltzer et al. 1984), supporting the hypothesis that an alteration in central 5-HT function may be relatively independent of state.

In contrast to prolactin responses, cortisol responses to 5-HTP challenge appear to be increased in patients with depression (or mania) compared with nondepressed control subjects (Meltzer et al. 1984). The increase in cortisol response to 5-HTP has been interpreted as reflecting increased sensitivity of postsynaptic 5-HT receptors, possibly in compensation for decreased presynaptic release of 5-HT. An inverse correlation between measures of CSF 5-HIAA and cortisol responses to 5-HTP in depressed patients, recently reported by Koyama and Meltzer (1986), provides some support for this possibility. However, why findings with tryptophan (or fenfluramine) challenge and 5-HTP challenge should be opposite in direction remains to be explained. It is possible that the prolactin and cortisol responses to these respective agents reflect the responsivity of different serotonergic subsystems defined either by brain location or by receptor subtype.

## Peripheral Indices of Serotonin

*Plasma Tryptophan.* In addition to the possibility of altered responsivity of central 5-HT receptors, diminished central 5-HT activity may be due to a decrease in the amount of 5-HT synthesized for release. Since the rate-limiting enzyme involved in the synthesis of 5-HT from tryptophan (tryptophan hydroxylase) is not normally saturated by its substrate, availability of tryptophan could play a role in the regulation of central presynaptic activity in some pathological conditions. Evidence of reduced availability of plasma free tryptophan, the species available to cross the blood-brain barrier, has been reported in approximately half of the more than 20 published studies of depressed patients (Moller et al. 1983). Even in the presence of normal plasma free tryptophan levels, availability of tryptophan may be influenced by the concentration of neutral amino acids competing for uptake across the blood-brain barrier (Fernstrom and Wurtman 1972). While changes in amino acid concentrations within the physiologic range may not meaningfully affect brain tryptophan levels, reduced free tryptophan/neutral amino acid ratios have been found in depressed patients compared with nondepressed control subjects in one study (Joseph et al. 1984), and these ratios have been found to correlate with severity of depression in another (DeMeyer et al. 1981). Finally, deficiencies of tryptophan metabolism have been reported in some (Koyama and Meltzer 1986; Smith and Stromgren 1981) but not all (Hoes et al. 1981; Moller et al. 1982) studies. This suggests that pharmacokinetic and dynamic factors may be associated with decreased availability and utilization of tryptophan in some depressed patients.

*Platelet Serotonin Uptake.* Because the human platelet appears to share many of the characteristics of central serotonergic neurons (Stahl 1977), indices of platelet 5-HT uptake are thought to reflect activity of central 5-HT uptake mechanisms. Reduced platelet 5-HT uptake has been reported in several studies of medication-free depressed patients (Meltzer and Lowy 1987), with uptake being most reduced in bipolar depressive patients in at least two studies (Goodnick and Meltzer 1984; Meltzer et al. 1981). Reduced 5-HT uptake into the platelet appears to be due to reduced numbers of 5-HT uptake sites ($V_{max}$) rather than reduced affinity of the recognition site ($K_m$) itself.

There is some evidence that reduced 5-HT uptake is a state-independent correlate of depression. $V_{max}$ does not change upon recovery in unipolar depressive patients (Coppen et al. 1978; Scott et al. 1979). Furthermore, reduced values have also been observed in manic patients (Meltzer et al. 1981). Although $V_{max}$ may be normal in drug-free recovered manic patients (Goodnick and Meltzer 1984), this may be due to a persistent effect of chronic lithium treatment, which is associated with an increase in $V_{max}$ values (Coppen et al. 1980). Finally, $V_{max}$ does not appear to correlate with severity of depression.

The significance of reduced $V_{max}$ in depression is unclear. Reduced numbers of central 5-HT uptake sites could lead to increased concentrations of synaptic 5-HT and increased 5-HT neurotransmission if there is not selective stimulation of inhibitory autoreceptors that would, in turn, decrease impulse-generated 5-HT synthesis and release. However, because uptake of 5-HT is the most efficient mechanism for deactivation and resupply of physiologically released 5-HT, this decrease in uptake could lead to decreased stores of presynaptic 5-HT. Alternatively, reduced 5-HT uptake sites could simply represent a compensatory response to a primary deficiency of presynaptic 5-HT.

*Platelet Tritiated-Imipramine Binding.* Like platelet 5-HT uptake, platelet tritiated-imipramine binding has been reported to be reduced in many, but not all, studies of depressed patients (reviewed in Briley 1985). Negative findings in some of these studies may be due to differences in the compositions of the patient samples, however, because reduced imipramine binding may only be characteristic of bipolar depressive patients or of unipolar depressive patients with a first-degree family history of depression (Lewis and McChesney 1985). Reduced imipramine binding is due to a decrease in the number of sites ($B_{max}$) that bind imipramine and not to the affinity of the site for imipramine. The imipramine binding site is allosterically related, but not identical, to the 5-HT uptake site (Briley 1985).

Unlike the case with $V_{max}$, however, decreased $B_{max}$ values appear to increase with recovery from depression and to increase with chronic treatment with some antidepressant agents (Aurora and Meltzer 1988). While the significance of reduced $B_{max}$ in depression is unclear, reduced $B_{max}$ (like $V_{max}$) may represent a compensatory response to a primary deficiency of presynaptic 5-HT.

## Serotonergic Treatment-Related Studies

Evidence for involvement of central 5-HT in the treatment of mood disorders comes from studies of the effect of 5-HT precursors (i.e., tryptophan, as reviewed by Cole et al. 1980; and 5-HTP, as reviewed by van Praag 1984), inhibition of 5-HT synthesis (Shopsin et al. 1975), and acute tryptophan depletion on mood in healthy subjects (Young et al. 1985) and depressed patients (Delgado et al. 1988). Evidence for the role of central 5-HT also comes from observations that nearly all effective antidepressants influence central 5-HT neurotransmission, either by increasing 5-HT output (i.e., through the direct and indirect effects of 5-HT reuptake inhibition) or by increasing 5-HT postsynaptic sensitivity (e.g., as observed with non–5-HT uptake inhibitors such as desipramine [Blier et al. 1987]).

The hypothesis that increased 5-HT neurotransmission alone is sufficient for antidepressant effects in depressed patients, however, is controversial. Evidence of increased central 5-HT neurotransmission with antidepressant treatment has been reported in depressed patients who do not respond to antidepressant treatment (Charney et al. 1984). Moreover, there is no evidence to suggest that selective serotonergic agents are more effective than other, less selective agents in the treatment of depression. Conceivably, patients may vary in the degree to which central 5-HT activity must be increased before antidepressant effects are observed. Alternatively, it is possible that interactions between central 5-HT and other neurotransmitters are altered in some patients in such a way that a pharmacologically induced increase in central 5-HT activity alone is inadequate to produce an antidepressant effect; this alteration perhaps is due to a disruption in the neurobiologic substrate upon which these agents act (Hsiao et al. 1987). Thus, although there is evidence for a role of central 5-HT in mediating the antidepressant response of many efficaciuos agents, the conclusion that alterations of 5-HT activity alone are accountable for the effects of antidepressant agents is premature at this time.

# SEROTONERGIC STUDIES OF PERSONALITY DISORDER

## Central Indices of Serotonin

*Lumbar CSF Serotonin Metabolite Concentration.* There are several studies that examine the concentration of lumbar CSF 5-HIAA in patients with one or more personality disorders. A strong inverse relationship between life history of physical aggression (i.e., Brown-Goodwin Assessment for Life History of Aggression [BGA]) and CSF 5-HIAA concentration ($r = -.78$, $N = 24$, $P < .001$) was first reported in a group of young military subjects, with a variety of DSM-II personality disorders, studied during a "fitness for duty" evaluation (Brown et al. 1979). Reduced CSF 5-HIAA concentrations (by median split) were associated with high aggression scores and with ultimate military discharge. Subjects with past history of suicide attempt were also characterized by significantly reduced concentrations of CSF 5-HIAA.

Cerebrospinal fluid 5-HIAA concentrations correlated inversely with BGA scores ($r = -.53$, $N = 12$, $P < .08$) and with the Psychopathic Deviance subscale of the Minnesota Multiphasic Personality Inventory (MMPI) ($r = -.77$, $N = 12$, $P < .005$) in a second study of 12 military subjects with DSM-III borderline personality disorder (Brown et al. 1982). While patients with low CSF 5-HIAA concentrations (by median split) had higher scores of Buss-Durkee "Motor Aggression," these differences did not reach statistical significance. CSF 5-HIAA was again found to be reduced in subjects with past history of suicide attempt, which in turn was associated with higher BGA scores. These findings suggested a trivariate relationship among aggression, suicide, and presumably, reduced central 5-HT function.

The relationship between violent aggressive behavior and reduced concentrations of CSF 5-HIAA has been a focus of research in this area since Åsberg et al. (1976b) first reported that depressed patients with a recent violent suicide attempt came from subjects in the "low" CSF 5-HIAA group. Suicide-attempting depressed patients using nonviolent methods had higher concentrations of CSF 5-HIAA—concentrations that were not lower than those among nonsuicide-attempting depressed patients. However, while many studies have replicated the finding of reduced CSF 5-HIAA concentrations in patients with a history of suicide attempt, not all studies have found a link between violence and suicide attempt (Åsberg et al. 1987).

Reduced CSF 5-HIAA concentrations have been reported specifically in violent offender subjects with a history of impulsive, nonpremeditated acts of violence (Linnoila et al. 1983c). Subjects in this group were diagnosed (DSM-II) as antisocial or as having explosive personality disorder, and they usually did not know the victims. Subjects who had committed deliberate aggressive acts were diagnosed as having passive personality disorder or paranoid personality disorder, and they demonstrated CSF 5-HIAA concentrations significantly higher than those of the impulsive violent offenders. Concentrations of CSF 5-HIAA in violent homocidal offenders have also been reported as reduced, but only in subjects who killed a loved one, presumably in a fit of rage (i.e., impulsive acts of violence) (Lidberg et al. 1985). Contract murderers who killed by profession (i.e., nonimpulsive acts of violence) did not demonstrate reduced concentrations of CSF 5-HIAA. Reduced CSF 5-HIAA concentrations have also been reported in impulsive arsonists (most of whom had comorbid borderline personality disorder by DSM-III criteria), in whom these concentrations correlated inversely with incidence of criminal acts other than arson (Virkkunen et al. 1987). Inverse correlations between a self-report inventory of outwardly directed hostility and CSF 5-HIAA concentrations in healthy volunteers (Roy et al. 1988) suggest that impulsive aggression may be associated with reduced central 5-HT activity even in nonpsychiatric subjects.

**Brain Serotonin Receptor Ligand Binding Studies.** There have been no studies examining brain receptor binding sites in patients who are explicitly described as personality disordered. However, there have been several postmortem studies of suicide victims who, in addition to possibly being depressed, may have had personality disorder. Reduced numbers of brain tritiated-imipramine binding sites in various regions of the brains of suicide victims, compared with accident victims, have been reported (Crow et al. 1984; Gross-Isseroff et al. 1989; Paul et al. 1984; Stanley et al. 1982), suggesting either diminished presynaptic output of 5-HT or reduced innervation of central 5-HT synapses. Conversely, increased numbers of $5-HT_2$ receptor binding sites in the frontal cortex of suicide victims have been reported (Aurora and Meltzer 1989; Ferrier et al. 1986; Mann et al. 1986; Stanley and Mann 1983). These latter findings have been interpreted as reflecting an increase in postsynaptic 5-HT receptor sites, possibly in response to diminished 5-HT output or innervation.

Cumulatively, these data suggest an abnormality in central 5-HT function in subjects who have committed suicide. While it is possible that pre- and postsynaptic factors vary in reciprocal fashion, the one study that examined the relationship between number of imipramine binding sites and number of 5-HT$_2$ receptor binding sites demonstrated only a modest, and nonsignificant, inverse correlation between the two ($r = -.42$, $P > .20$) (Stanley and Mann 1983). These results suggest either that these indices of central 5-HT function are not closely regulated together or that there is a dissociation between the functioning of the pre- and postsynaptic neurons in suicidal patients. It is possible that the degree to which the 5-HT postsynaptic receptor can increase in number (or sensitivity) is limited because of a more primary defect in receptor regulatory mechanisms, in which case 5-HT synaptic function would be poorly compensated and diminished central 5-HT activity would result. However, it is impossible to test this hypothesis without assessing pre- and postsynaptic 5-HT function simultaneously.

*Serotonin Pharmacochallenge Studies.* The use of pharmacologic challenge, as reviewed above, offers the opportunity to test the "net" activity of the central serotonergic system. In order to examine central 5-HT function in relation to aggression and impulsivity, a pharmacologic challenge with fenfluramine was administered in our laboratory to patients with DSM-III personality disorder.

In a recently published study (Coccaro et al. 1989), peak delta prolactin responses to fenfluramine were reduced in 20 patients with a variety of DSM-III personality disorders when compared with 14 healthy control subjects of similar age (Figure 4-2). Prolactin responses to fenfluramine were inversely correlated with clinician- and self-rated measures of aggression and impulsivity in the overall patient group. Significant negative correlations were found between peak delta prolactin responses to fenfluramine and (a) Brown-Goodwin "Aggression" (i.e., BGA) ($r = -.57$, $P < .012$), (b) Buss-Durkee "Motor Aggression" ($r = -.52$, $P < .019$), and (c) Barratt Total "Impulsiveness" ($r = -.48$, $P < .03$). While a negative correlation was found with the Psychopathic Deviance subscale of the MMPI ($r = -.33$, $P < .16$), this correlation was not statistically significant. Correlational analysis of data from other, nonaggressive/nonimpulsive, dimensional items from the remaining 12 MMPI subscales, the Spielberger State-Trait Anxiety Inventory, and the Zuckerman Sensation Seeking Scale, did not yield statistically significant correlations with prolactin responses to fenfluramine, suggesting that the prolactin responses to fenfluramine did not correlate nonspecifically with behavioral, or personality, dimensions that are reflective of anxiety, sensation seeking, or of other areas of general psychopathology.

Examination of the subscales of the self-reported assessments that

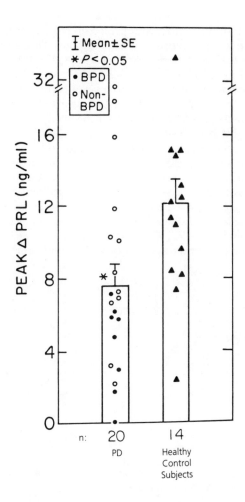

**Figure 4-2.** Peak change in prolactin (peak ΔPRL [ng/ml]) responses to fenfluramine challenge in 20 male drug-free DSM-III personality disorder (PD) patients (circles; aged 37.2 ± 7.8 years) and 14 male drug-free healthy control volunteers (closed triangles) of similar age (39.3 ± 7.9 years). Closed circles represent PD patients meeting DSM-III criteria for borderline personality disorder (BPD); open circles represent PD patients who did not meet DSM-III criteria for BPD. Reproduced from Coccaro et al., *Archives of General Psychiatry* 46:587–599, 1989, with permission of the American Medical Association. Copyright 1989, American Medical Association.

correlated significantly with prolactin responses to fenfluramine revealed that the subscales relating to Buss-Durkee "Assault" ($r = -.65$, $P = .002$) and "Irritability" ($r = -.68$, $P < .002$), and to Barratt "Motor Impulsiveness" ($r = -.54$, $P < .013$), were the specific dimensions accounting for these inverse correlations with prolactin responses to fenfluramine. These three subscales were highly correlated with the BGA ($r$ values from .58 to .74; $P < .01$ to .0001), suggesting that each of these measures reflects overlapping dimensions of behavior related to irritable, impulsive motoric aggression. Further analysis revealed that of these four measures 59% of the variance in prolactin responses to fenfluramine could be accounted for by Buss-Durkee "Assault" and "Irritability" scores alone (multiple $R = .77$, $F(2,17) = 12.54$, $P < .001$). The sum score of these two measures alone correlated highly with the prolactin responses to fenfluramine ($r = -.77$, $P = .0002$) (Figure 4-3, left).

Reduced prolactin responses to fenfluramine in the overall patient

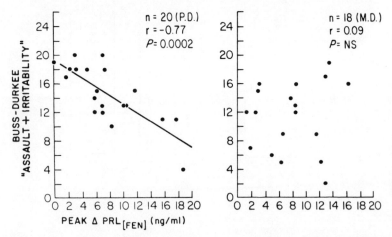

**Figure 4-3.** Left: Correlation between peak change in prolactin (peak ΔPRL [ng/ml]) responses to fenfluramine and the sum score of Buss-Durkee "Assault" and "Irritability" subscales for the 20 patients with primary personality disorder (PD). Beta values from multiple regression analysis (see text) were $-.42$ ($P < .03$) for Buss-Durkee "Assault" and $-.48$ ($P < .015$) for Buss-Durkee "Irritability." Right: Correlation between peak delta prolactin responses to fenfluramine and the sum score of Buss-Durkee "Assault" and "Irritability" subscales for 18 patients with primary mood disorder (MD; acute = 8, remitted = 10).

group were largely due to the reduced responses in patients meeting DSM-III criteria for borderline personality disorder (Figures 4-2 and 4-4 [top]). Prolactin responses to fenfluramine among patients meeting DSM-III criteria for a personality disorder other than borderline personality disorder were significantly greater than those among patients with borderline personality disorder and not significantly different from those observed among the healthy control subjects. Similar examination of patients with and without other specific Axis II disorders (e.g., schizotypal, paranoid, histrionic) revealed that only patients with borderline personality disorder displayed significantly reduced prolactin responses to fenfluramine.

However, presence of a past history of suicide attempt and/or of presence of a past history of alcohol abuse, both of which may be associated with central serotonergic dysfunction (see Chapter 8), were also associated with reduced prolactin responses to fenfluramine (Figure 4-4, center and bottom). Since a past history of suicide attempt and/or alcohol abuse is frequently associated with the diagnosis of borderline personality disorder, it is possible that a related factor, such as presence of impulsive-aggressive traits, may account for these findings. In this sample, patients with either borderline personality disorder, past history of suicide attempt, or alcohol abuse systematically scored higher on measures related to impulsive aggression than did patients without these characteristics. Analysis of covariance revealed that the group differences in prolactin responses to fenfluramine between patients with and without borderline personality disorder, or with and without a past history of suicide attempt or alcohol abuse, did not remain significant after the influence of these behavioral dimensions (i.e., Buss-Durkee "Assault" and "Irritability" scores) was removed (Figure 4-4). This finding suggests that despite the categorical group differences observed, the most powerful behavioral correlate of the prolactin response to fenfluramine, and perhaps of "net" central serotonergic function, is that described by "irritable" impulsive aggression. This hypothesis is consistent with the findings of numerous preclinical studies implicating central 5-HT in the mediation of "irritable" aggression (reviewed in DePue and Spoont 1986; Soubrie 1986).

### Peripheral Indices of Seotonin

While there are at present no studies utilizing specific peripheral indices of serotonergic function in patients with DSM-III-R personality disorder, an inverse correlation between numbers of platelet tritiated-imipramine binding sites and measures of aggressivity in adolescents with the diagnosis of conduct disorder has been reported

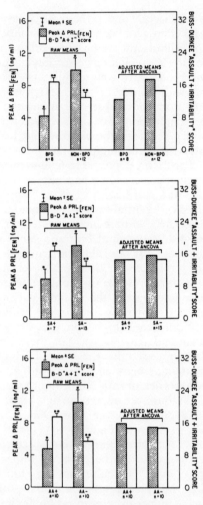

**Figure 4-4.**    Peak change in prolactin (peak $\Delta$PRL [ng/ml]) responses to fenfluramine and the sum score of Buss-Durkee "Assault" and "Irritability" subscales among DSM-III personality disorder patients, with and without DSM-III borderline personality disorder (BPD $+/-$; top), past history of suicide attempt (SA $+/-$; center), or alcohol abuse (AA $+/-$; bottom). Left side of panels represents raw means for both variables. Right side of panels represents adjusted means for peak delta prolactin responses to fenfluramine after ANCOVA (sum score of Buss-Durkee "Assault" and "Irritability" subscales as covariate).

(Stoff et al. 1987). Conduct disorder is a behavioral disorder of childhood and adolescence that, according to DSM-III-R criteria, must have been present in the adolescent histories of adult patients who otherwise meet criteria for antisocial personality disorder. Thus it is possible that this reported relationship—similar to that observed between lumbar CSF 5-HIAA concentrations or the prolactin response to fenfluramine and aggression in adults—may also be present in antisocial adults.

Similarly, reduced plasma levels of tryptophan have been reported in alcoholic patients with a history of aggressive or suicidal behavior compared to alcoholic patients without this history (Branchey et al. 1984). Although diagnoses of personality disorder were not reported, it is highly unlikely that this group of chronic alcoholic patients were free of Axis II disorders.

Data from studies of whole-blood 5-HT are contradictory. While an inverse relationship between whole-blood 5-HT and impulsive-aggressive behavior has been reported in mentally retarded patients with hyperactivity and impulsive-aggressive behaviors (Greenberg and Coleman 1976), a positive relationship between whole-blood 5-HT and ratings of conduct disorder behavior was recently reported in adolescents (Pliszka et al. 1988).

## Serotonin Treatment-Related Studies

Despite a growing literature suggesting that behavioral dimensions related to "irritable" impulsive aggression are inversely related to central serotonergic function, few studies have tested the hypothesis that these behaviors may be diminished with treatments that putatively increase central serotonergic activity. To date, the only well-designed study in this regard is a placebo-controlled crossover trial by Sheard et al. (1976), who reported that lithium treatment was associated with a significant reduction of physically aggressive behaviors in a group of prison inmates. Although (DSM-II) diagnoses of personality disorder were not reported, it is likely that many of these subjects would meet the DSM-III-R criteria for antisocial personality disorder. Other studies examining the effect of lithium on aggressive behavior in a variety of diagnostic groups have reported similar results (reviewed in Wickham and Reed 1987). Coupled with data from animal and clinical psychobiologic studies (reviewed above), the antiaggressive action of lithium has been interpreted as reflecting a possible serotonergic mechanism. However, lithium may also diminish catecholaminergic function (Linnoila et al. 1983a, 1983b), and data suggestive of a positive relationship between noradrenergic (Brown et al. 1979) and/or dopaminergic (Senault 1970) activity and aggressive behaviors, have been reported. Thus, it is not known if the antiaggressive activity of lithium is mediated by serotonergic mechanisms, catecholaminergic mechanisms, or by a balance between the two. Given the notion of a balance between neuronal systems that mediate behavioral arousal and inhibition (DePue and Spoont 1986), it is possible that serotonergic and anticatecholaminergic actions of lithium may mediate this effect.

Other pharmacologic interventions reported to diminish impulsive-aggressive behavior in psychiatric patients include treatment with low-dose neuroleptics (Campbell et al. 1984; Montgomery and Montgomery 1982; Soloff et al. 1986), high-dose nonselective beta-blockers (Elliot 1977; Sorgi et al. 1986; Yudofsky et al. 1981), and carbamazepine (Cowdry and Gardner 1988). While none of these agents appear to have a clear central serotonergic mode of action, there is evidence suggesting that agents from each of these pharmaceutical classes have some effect on central 5-HT activity. For example, many neuroleptics display moderate affinity for $5-HT_2$ receptors (Ortmann et al. 1982). Nonselective beta-blockers display high affinity for $5-HT_1$ receptors (Hoyer 1988) and may act as $5-HT_1$ agonists (Hjorth and Carlsson 1986) at high doses. Finally, carbamazepine has been reported to increase plasma levels of tryptophan (Pratt et al. 1984), which could, in turn, be converted into 5-HT in the brain. Whether the efficacy of these agents in the treatment of impulsive aggression is due to possible effects on central 5-HT neurotransmission remains to be tested.

Presently, there are at least two 5-HT–specific agents available for clinical trials in the treatment of impulsive aggression. These agents include fluoxetine, a selective 5-HT uptake inhibitor antidepressant, and buspirone, an atypical anxiolytic with high affinity for central $5-HT_{1A}$ receptors (Hoyer 1988). Animal studies have shown these agents, and others in their class, to be effective in diminishing aggression (Berzsenyi et al. 1983; McMillen et al. 1987; Molina et al. 1986; Olivier et al. 1984) and thus support a rationale for use of these agents in clinical trials. While clinical reports already suggest that both agents may be useful in this regard, double-blinded, placebo-controlled trials will be necessary before clinicians should routinely include these agents in the treatment of the impulsive-aggressive individual. Other 5-HT–specific agents currently under development, and potentially available for similar efficacy trials, include the 5-HT–specific uptake inhibitors sertraline and fluvoxamine, and the $5-HT_{1A}$ and $5-HT_{1A-B}$ agonists gepirone and eltoprazine, respectively.

# INTERACTIONS BETWEEN SEROTONIN AND MOOD OR PERSONALITY DISORDER

A clear understanding of the role of central 5-HT in either mood or personality disorder awaits further research. As reviewed above, there is evidence that 5-HT function may be reduced in patients with major depression, in patients with or without history of suicide attempt, and in patients with personality disorder in whom features of "irritable" impulsive aggression are prominent. However, it is unclear if abnormalities of 5-HT function correlate with these specific behaviors in patients with mood or personality disorder.

In our studies (Coccaro et al. 1989) we found a relationship between reduced prolactin response to fenfluramine and history of suicide attempt in mood disorder and personality disorder patients regardless of diagnosis. However, while a trivariate relationship among reduced prolactin response to fenfluramine, suicide attempt, and "irritable" impulsive aggression could be demonstrated for patients with primary personality disorder (Figure 4-4, center), no such relationship could be demonstrated for patients with primary mood disorder (Figure 4-3, right). Increased aggressive and/or antisocial behavior in suicidal depressed patients with low CSF 5-HIAA concentrations has been reported (Rydin et al. 1982; van Praag 1986); however, it is unknown how these patients compare to those in our study who were prospectively dichotomized into two groups— those patients with primary mood disorder and those with primary personality disorder. It is possible that patients from these previous studies would have fallen into our primary personality disorder group, most of whom also had histories of major depression at the time of the study or in the past. Associations between low CSF 5-HIAA concentrations and high scores on measures of impulsivity, sensation seeking, and psychopathy in nondepressed patients only (Schalling 1986), as well as the absence of a relationship between low CSF 5-HIAA concentration and suicide attempt in bipolar depressive patients (Åsberg et al. 1987), support the possibility that indices of 5-HT activity may not correlate with behaviors related to outwardly directed impulsive aggression in some psychiatric populations. This suggests the possibility that presence of primary mood disorder modifies the behavioral expression of reduced central 5-HT activity away from outwardly directed impulsive aggression to self-directed aggression (i.e., suicide), particularly in patients with a relative absence of clinically significant personality disturbance. The source of this variance is unknown but may lie in altered relationships among different neurotransmitter systems (e.g., catecholaminergic and

serotonergic) (Hsiao et al. 1987) and/or in the differential functioning of various neuronal systems subserved by 5-HT.

## CONCLUSIONS

Central and peripheral indices of central serotonergic function have
been shown to be abnormal in a variety of patients with mood and/or
personality disorder. Although these abnormalities are not specific for
major mood disorder per se, these indices may have the potential to
reflect behavioral correlates (e.g., categorical history of and vulnerability to mood disturbance, suicide attempt, and "irritable" impulsive aggression) of aberrant central serotonergic function as may
exist in a variety of major psychiatric disorders. Further work is needed
to more fully understand the nature of these abnormalities in regard
to 5-HT synaptic physiology, as well as interactions between central
5-HT and other neurotransmitter systems, in order to elucidate the
mechanisms by which central 5-HT regulates mood and impulse
control in patients with psychiatric disorders.

## REFERENCES

Agren H: Symptom patterns in unipolar and bipolar depression correlating
with monoamine metabolites in the cerebrospinal fluid: general patterns.
Psychiatry Res 3:211–224, 1980

Åsberg M, Thorén P, Träskman L, et al: "Serotonin depression"—a
biochemical subgroup within the affective disorders? Science 191:478–
480, 1976a

Åsberg M, Träskman L, Thorén P: 5-HIAA in the cerebrospinal fluid: a
biochemical suicide predictor? Arch Gen Psychiatry 33:1193–1197,
1976b

Åsberg M, Bertilsson L, Martenssen B, et al: CSF monoamine metabolites
in melancholia. Acta Psychiatr Scand 69:201–219, 1984

Åsberg M, Schalling D, Träskman-Bendz L, et al: Psychobiology of suicide,
impulsivity, and related phenomena, in Psychopharmacology: The Third
Generation of Progress. Edited by Meltzer HY. New York, Raven Press,
1987, pp 655–668

Asnis GM, Eisenberg J, van Praag HM, et al: The neuroendocrine response
to fenfluramine in depressives and normal controls. Biol Psychiatry
24:117–120, 1988

Aurora RC, Meltzer HY: Effect of desipramine treatment on 3-H-imipramine
binding in the blood platelets of depressed patients. Biol Psychiatry
23:397–404, 1988

Aurora RC, Meltzer HY: Serotonergic measures in the brains of suicide victims: 5-HT-2 binding sites in the frontal cortex of suicide victims and control subjects. Am J Psychiatry 146:730–736, 1989

Berzsenyi P, Galateo E, Valzelli L: Fluoxetine activity on muricidal aggression induced in rats by *p*-chloroamphetamine. Aggressive Behavior 9:333–338, 1983

Blier P, de Montigny C, Chaput Y: Modifications of the serotonin system by antidepressant treatments: implications for the therapeutic response in major depression. J Clin Psychopharmacol 7:24S–35S, 1987

Branchey L, Branchey M, Shaw S, et al: Depression, suicide, and aggression in alcoholics and their relationship to plasma amino acids. Psychiatry Res 12:219–226, 1984

Briley M: Imipramine binding: its relationship with serotonin, in Neuropharmacology of Serotonin. Edited by Green AR. Oxford, UK, Oxford University Press, 1985, pp 50–78

Brown GL, Goodwin FK, Ballenger JC, et al: Aggression in humans correlates with cerebrospinal fluid amine metabolites. Psychiatry Res 1:131–139, 1979

Brown GL, Ebert MH, Goyer PF, et al: Aggression, suicide, and serotonin: relationships to CSF amine metabolites. Am J Psychiatry 139:741–746, 1982

Campbell M, Small A, Green WH, et al: Behavioral efficacy of haloperidol and lithium carbonate: a comparison in hospitalized aggressive children with conduct disorder. Arch Gen Psychiatry 41:650–656, 1984

Charney DS, Heninger GR, Sternberg DE: Serotonin function and mechanism of action of antidepressant treatment. Arch Gen Psychiatry 41:359–365, 1984

Coccaro EF, Siever LJ, Klar HM, et al: Serotonergic studies in affective and personality disorder patients: correlates with suicidal and impulsive aggressive behavior. Arch Gen Psychiatry 46:587–599, 1989

Cole JO, Hartmann E, Brigham P: L-Tryptophan: clinical studies, in Psychopharmacology Update. Edited by Cole JO. Lexington, MA, Collamore Press, 1980, pp 119–148

Coppen A, Swade C, Wood K: Platelet 5-hydroxytryptamine accumulation in depressive illness. Clin Chim Acta 87:165–168, 1978

Coppen A, Swade C, Wood K: Lithium restores abnormal platelet 5-HT transport in patients with affective disorders. Br J Psychiatry 136:235–238, 1980

Cowdry RW, Gardner DL: Pharmacotherapy of borderline personality disorder: alprazolam, carbamazepine, trifluoperazine, and tranylcypromine. Arch Gen Psychiatry 45:111–119, 1988

Cowen PJ, Charig EM: Neuroendocrine responses to intravenous tryptophan in major depression. Arch Gen Psychiatry 44:958–966, 1987

Crow TJ, Cross AJ, Cooper SJ, et al: Neurotransmitter receptors and monoamine metabolites in brains of patients with Alzheimer-type dementia and depression, and suicides. Neuropharmacology 23:1561–1569, 1984

Davis KL, Hollister LE, Mathe AA, et al: Neuroendocrine and neurochemical measurements in depression. Am J Psychiatry 138:1555–1562, 1981

Delgado P, Charney DS, Price LH, et al: Behavioral effects of acute tryptophan-depletion in depressed patients. Abstract of paper presented at 27th annual meeting of the American College of Neuropsychopharmacology, San Juan, Puerto Rico, December 1988, p 77

DeMeyer MK, Shea PA, Hendrie HC, et al: Plasma tryptophan and five other amino acids in depressed and normal subjects. Arch Gen Psychiatry 38:642–646, 1981

DePue RA, Spoont MR: Conceptualizing a serotonin trait: a behavioral dimension of constraint. Ann NY Acad Sci 487:47–62, 1986

Elliot FA: Propranolol for the control of belligerent behavior following acute brain damage. Ann Neurol 1:489–491, 1977

Fernstrom JD, Wurtman RJ: Brain serotonin content: physiological regulation by plasma neutral amino acids. Science 178:414–416, 1972

Ferrier IN, McKeith IG, Cross AJ, et al: Postmortem neurochemical studies in depression. Ann NY Acad Sci 487:128–142, 1986

Gibbons RD, Davis JM: Consistent evidence for a biological subtype of depression characterized by low CSF 5-HIAA monoamine levels. Acta Psychiatr Scand 74:8–12, 1986

Goodnick PJ, Meltzer HY: Neurochemical changes during discontinuation of lithium prophylaxis. II. Alterations in serotonin function. Biol Psychiatry 19:891–898, 1984

Greenberg AS, Coleman M: Depressed 5-hydroxyindole levels associated with hyperactive and aggressive behavior. Arch Gen Psychiatry 33:331–336, 1976

Gross-Isseroff R, Israeli M, Biegon A: Autoradiographic analysis of tritiated imipramine binding in the human brain post-mortem: effects of suicide. Arch Gen Psychiatry 46:237–241, 1989

Heninger GR, Charney DS, Sternberg DE: Serotonergic function in depression. Arch Gen Psychiatry 41:398–402, 1984

Hjorth S, Carlsson A: Is pindolol a mixed agonist/antagonist at central serotonin (5-HT) receptors? Eur J Pharmacol 129:131–138, 1986

Hoes MJAJM, Loeffen T, Vree TB: Kinetics of L-tryptophan in depressed patients: a possible correlation between plasma concentrations of L-tryptophan and some psychiatric rating scales. Psychopharmacology (Berlin) 75:350–353, 1981

Hoyer D: Functional correlates of serotonin-1 recognition sites. J Recept Res 8:59–81, 1988

Hsiao JK, Agren H, Bartko JJ, et al: Monoamine neurotransmitter interactions and the prediction of antidepressant response. Arch Gen Psychiatry 44:1078–1083,1987

Invernizzi R, Kmieciak-Kolada K, Samanin R: Is receptor activation involved in the mechanism by which (+)-fenfluramine and (+)-norfenfluramine deplete 5-hydroxytryptophan in the rat brain? Br J Pharmacol 75:525–530, 1982

Joseph M, Brewerton T, Reus V, et al: Plasma L-tryptophan/neutral amino acid ratio and dexamethasone suppression in depression. Psychiatry Res 11:185–192, 1984

Koyama T, Meltzer HY: A biochemical and neuroendocrine study of the serotonergic system in depression, in New Results in Depression Research. Edited by Hippius H, Klerman GL, Matussek N. Berlin, Springer-Verlag, 1986, pp 169–188

Kuhn DM, Wolf W, Youdim MBH: Serotonin neurochemistry revisited: a new look at some old axioms. Neurochem Int 2:141–154, 1986

Lewis DA, McChesney C: Tritiated imipramine binding distinguishes among subtypes of depression. Arch Gen Psychiatry 42:485–488, 1985

Lidberg L, Tuck JR, Åsberg M, et al: Homocide, suicide and CSF 5-HIAA. Acta Psychiatr Scand 71:230–236, 1985

Linnoila M, Karoum F, Potter WZ: Effects of antidepressant treatments on dopamine turnover in depressed patients. Arch Gen Psychiatry 40:1015–1017, 1983a

Linnoila M, Karoum F, Rosenthal N, et al: Electroconvulsive treatment and lithium carbonate. Their effects on norepinephrine metabolism in patients with primary, major depressions. Arch Gen Psychiatry 40:677–680, 1983b

Linnoila M, Virkkunen M, Scheinin M, et al: Low cerebrospinal fluid

5-hydroxyindoleacetic acid concentration differentiates impulsive from nonimpulsive violent behavior. Life Sci 33:2609–2614, 1983c

Lopez-Ibor JJ, Saiz-Ruiz J, Moral Iglesias L: The fenfluramine challenge test in the affective spectrum: a possible marker of endogeneity and severity. Pharmacopsychiatry 21:9–14, 1988

Mann JJ, Stanley M, McBride PA, et al: Increased serotonin-2 and beta-adrenergic receptor binding in the frontal cortices of suicide victims. Arch Gen Psychiatry 43:954–959, 1986

McMillen BA, Scott SM, Williams HL, et al: Effect of geperone, an aryl-piperazine anxiolytic drug, on aggressive behavior and brain monoaminergic neurotransmission. NS Arch Pharmacol 335:454–464, 1987

Meltzer HY, Lowy MT: The serotonin hypothesis of depression, in Psychopharmacology: The Third Generation of Progress. Edited by Meltzer HY. New York, Raven Press, 1987, pp 513–526

Meltzer HY, Aurora RC, Baber R, et al: Serotonin uptake in blood platelets of psychiatric patients. Arch Gen Psychiatry 38:1322–1326, 1981

Meltzer HY, Umberkomen-Wiita B, Robertson A, et al: Effect of 5-hydroxytryptophan on serum cortisol levels in major affective disorders. I. Enhanced response in depression and mania. Arch Gen Psychiatry 41:366–374, 1984

Molina VA, Gobaille S, Mandel P: Effects of serotonin-mimetic drugs on mouse-killing behavior. Aggressive Behav 12:201–211, 1986

Moller SE, Kirk L, Honore P: Tryptophan tolerance and metabolism in endogenous depression. Psychopharmacology (Berlin) 76:79–83, 1982

Moller SE, Kirk L, Brandrup E, et al: Plasma tryptophan in depression. Biol Psychiatry 10:30–46, 1983

Montgomery SA, Montgomery D: Pharmacological prevention of suicidal behavior. J Affective Disord 4:291–298, 1982

Olivier B, Van Aken M, Jaarsma I, et al: Behavioral effects of psychoactive drugs on agonistic behavior of male territorial rats (resident-intruder model). Prog Clin Biol Res 167:137–156, 1984

Ortmann R, Bischoff S, Radeke E, et al: Correlations between different measures of antiserotonin activity of drugs: study with neuroleptics and serotonin receptor blockers. NS Arch Pharmacol 321:265–270, 1982

Paul SM, Rehavi M, Skolnick P, et al: High affinity binding of antidepressants to a biogenic amine transport site in human brain and platelet: studies in depression, in Neurobiology of Mood Disorders. Edited by Post RM, Ballenger JC. Baltimore, MD, Williams & Wilkins, 1984, pp 845–953

Pliszka SR, Rogeness GA, Renner P, et al: Plasma neurochemistry in juvenile offenders. Am Acad Child Adolesc Psychiatry 27:588–594, 1988

Pratt JA, Jenner P, Johnson AL, et al: Anticonvulsant drugs alter plasma tryptophan concentrations in epileptic patients: implications for anti-epileptic action and mental function. J Neurol Neurosurg Psychiatry 47:1131–1133, 1984

Rothpearl AB, Mohs RC, Davis KL: Statistical power in biological psychiatry. Psychiatry Res 5:257–266, 1981

Roy A, Pickar D, Linnoila M, et al: Cerebrospinal fluid monoamine and monoamine metabolite concentrations in melancholia. Psychiatry Res 15:281–292, 1985

Roy A, Adinoff B, Linnoila M: Acting out hostility in normal volunteers: negative correlation with CSF 5-HIAA levels. Psychiatry Res 24:187–194, 1988

Rydin E, Schalling D, Åsberg M: Rorschach ratings in depressed and suicidal patients with low levels of 5-hydroxyindoleacetic acid in cerebrospinal fluid. Psychiatry Res 7:229–243, 1982

Schalling D: The involvement of serotonergic mechanisms in anxiety and impulsivity in humans (an open peer commentary on "Reconciling the role of central serotonin neurons in human and animal behavior" by P Soubrie). Behav Brain Sci 9:343–344, 1986

Scott M, Reading HW, Loudon JB: Studies on human blood platelets in affective disorders. Psychopharmacology (Berlin) 60:131–135, 1979

Senault B: Comportement d'aggressive intraspecifique indivit par l'apo-morphine chez le rat. Psychopharmacologia 18:271–287, 1970

Sheard MH, Marini JL, Bridges CI, et al: The effect of lithium on impulsive aggressive behavior in man. Am J Psychiatry 133:1409–1413, 1976

Shopsin B, Friedman E, Goldstein M, et al: The use of synthesis inhibitors in defining a role for biogenetic amines during imipramine treatment in depressed patients. Psychopharmacol Commun 1:239–249, 1975

Siever LJ, Murphy DL, Slater S, et al: Plasma prolactin changes following fenfluramine in depressed patients compared to controls: an evaluation of central serotonergic responsivity in depression. Life Sci 34:1029–1039, 1984

Smith DF, Stromgren LS: Influence of unilateral ECT on tryptophan metabolism in endogenous depression. Pharmacopsychiatry 14:135–138, 1981

Soloff PH, George A, Nathan RS, et al: Progress in the pharmacotherapy of borderline disorders: a double-blind study of amitriptyline, haloperidol, and placebo. Arch Gen Psychiatry 43:691–697, 1986

Sorgi PJ, Ratey JJ, Polakoff S: Beta-adrenergic blockers for the control of aggressive behaviors in patients with chronic schizophrenia. Am J Psychiatry 143:775–776, 1986

Soubrie P: Reconciling the role of central serotonin neurons in human and animal behavior. Behav Brain Sci 9:319–364, 1986

Stahl SM: The human platelet. Arch Gen Psychiatry 34:509–516, 1977

Stanley MS, Mann JJ: Increased serotonin-2 binding sites in frontal cortex of suicide victims. Lancet 1:214–216, 1983

Stanley MS, Viggilio J, Gershon S, et al: Tritiated imipramine binding sites are decreased in the frontal cortex of suicides. Science 216:1337–1339, 1982

Stoff DM, Pollack L, Vitiello B, et al: Reduction of 3-H-imipramine binding sites on platelets of conduct-disordered children. Neuropsychopharmacology 1:55–62, 1987

Träskman-Bendz L, Åsberg M, Bertilsson L, et al: CSF monoamine metabolites of depressed patients during illness and after recovery. Acta Psychiatr Scand 69:333–342, 1984

van Praag HM: Studies in the mechanism of action of serotonin precursors in depression. Psychopharmacol Bull 20:599–602, 1984

van Praag HM: (Auto)aggression and CSF 5-HIAA in depression and schizophrenia. Psychopharmacol Bull 22:669–673, 1986

van Praag HM, de Haan S: Central serotonin metabolism and frequency of depression. Psychiatry Res 1:219–224, 1979

van Praag HM, de Haan S: Depression vulnerability and 5-HTP prophylaxis. Psychiatry Res 3:75–83, 1980

Virkkunen M, Nuutila A, Goodwin FK, et al: Cerebrospinal fluid monoamine metabolite levels in male arsonists. Arch Gen Psychiatry 44:241–247, 1987

Voyt M: Biology of Serotonergic Neurotransmission. Chichester, UK, John Wiley, 1982

Wickham EA, Reed JV: Lithium for the control of aggressive and self-mutilating behavior. Int Clin Psychopharmacol 2:181–190,1987

Young SN, Smith S, Pihl RO, et al: Tryptophan depletion causes a rapid lowering of mood in normal males. Psychopharmacology (Berlin) 87:173–177, 1985

Yudofsky S, Williams D, Gorman J: Propranolol in the treatment of rage and violent behavior in patients with chronic brain syndromes. Am J Psychiatry 138:218–220, 1981

# Chapter 5

## *Serotonin in Obsessive-Compulsive Disorder*

Joseph Zohar, M.D.
Dennis L. Murphy, M.D.
Rachel C. Zohar-Kadouch, M.D.
Michelle A. T. Pato, M.D.
Krystina M. Wozniak, M.D.
Thomas R. Insel, M.D.

# Chapter 5

## *Serotonin in Obsessive-Compulsive Disorder*

Obsessive-compulsive disorder (OCD) is a syndrome characterized by persistent, unwanted thoughts or impulses (obsessions), and repetitive, ritualistic behaviors that the person feels driven to perform (compulsions). Although both obsessions and compulsions are found in a variety of psychiatric disorders, as well as in normal mental life, OCD is distinguished by two cardinal features. First, the symptoms are ego-dystonic; the individual attempts to ignore or suppress the obsessions and/or the compulsions and recognizes that these preoccupations are excessive or unreasonable. Second, the obsessions and/or the compulsions cause marked distress, are time consuming (i.e., take more than 1 hour each day), and lead to significant interference in functioning.

Unlike patients with major psychotic syndromes, patients with OCD retain some insight about their symptoms. Although in some cases the behavior of OCD patients may appear to be extremely bizarre or paranoid, the idea (obsession) or ritual (compulsion) is recognized by patients with OCD as internal as opposed to the psychotic "idea of influence." Other behaviors in which people engage excessively, and with a sense of compulsion, such as pathologic gambling, overeating, alcohol or drug abuse, and hypersexuality, are also distinguished from true compulsions because to some degree these behaviors are experienced as pleasurable, while compulsions are inherently not pleasurable.

The OCD syndrome needs to be differentiated from obsessive-compulsive personality disorder. Individuals with this personality disorder exhibit perfectionism, orderliness, and rigidity, often with an inability to grasp "the big picture." These traits are ego-syntonic (i.e., the individual does not attempt to ignore or suppress them). Epidemiological evidence reveals that a substantial number of patients with OCD do not exhibit premorbid obsessive-compulsive personality traits (Rasmussen and Tsuang 1984).

101

OCD presents in several forms (Insel 1984). The most common symptoms are obsessions about dirt or contamination (Akhtar et al. 1975; Rachman and Hodgson 1980) accompanied by ritualistic washing. Paradoxically, some patients with these symptoms are quite slovenly. "Checkers" are obsessed with a doubt, usually a doubt tinged with guilt, or a need for symmetry; they will check repetitively to reassure themselves. Typically, they fear that the doors have been unlocked, that a bump on the road is a body, or that the book they are trying to read is not "in the right" position. Their checking, instead of resolving the uncertainty, often contributes to an even greater doubt.

Approximately 25% of OCD patients have pure obsessions without compulsive rituals (Akhtar et al. 1975; Welner et al. 1976). These patients have repetitive intrusive thoughts or impulses, usually aggressive (e.g., "I don't use knives because I am afraid that I might stab my daughter") or sexual (e.g., "People may think I am homosexual!"), and always reprehensible. Primary obsessional slowness is a rare yet very disabling form of the syndrome (Rachman 1974). These patients will spend hours each day on the simplest acts of personal hygiene and may become housebound.

## PREVALENCE OF OBSESSIVE-COMPULSIVE DISORDER

The prevalence of OCD in the general population, although traditionally thought to be low (0.05%) (Woodruff and Pitts 1969), has recently been reported to approximate 2%, according to the recent National Institute of Mental Health (NIMH)–sponsored Epidemiologic Catchment Area (ECA) Survey (Karno et al. 1988; Robins et al. 1984). Another study that involved direct interviews of children in the general population found a prevalence of 0.5% (Flament et al. 1989). Because only one-third of adult cases of OCD report onset in childhood, and because OCD is a chronic disorder (Insel 1984), this finding would suggest a projected prevalence of OCD of 1.5% in the general population, in agreement with the ECA results.

## CLOMIPRAMINE AND OBSESSIVE-COMPULSIVE DISORDER

Until recently, the outlook regarding pharmacological treatment for OCD was pessimistic. Salzman and Thaler (1981), in their review of the pre-1978 literature, concluded that "with regard to drugs . . . there is neither convincing nor suggestive evidence that more can be accomplished than the relief of some anxiety—at a cost that often outweighs the potential benefits" (p. 293).

Over 20 years ago, Renynghe de Voxurie (1968) reported that the tricyclic antidepressant clomipramine reduced the obsessional symptoms in 10 of his 15 obsessive patients. During the 1970s, a series of confirmatory but uncontrolled studies extended this initial observation to other patients with primary OCD (Ananth et al. 1979; Rack 1977; Yaryura-Tobias and Neziroglu 1975).

More recently, a group of carefully controlled double-blind studies using clomipramine have been published (Table 5-1). These controlled studies, in which clomipramine was administered collectively to 116 patients, have yielded fairly consistent results. In four studies (Flament et al. 1985; Marks et al. 1980; Montgomery 1980; Thorén et al. 1980b) clomipramine was more effective than placebo in reducing OCD symptomatology. In several parallel studies, clomipramine has been compared with nortriptyline (Thorén et al. 1980b), amitriptyline (Ananth et al. 1981), and imipramine (Volavka et al. 1985). In a small sample of patients, clomipramine was slightly better than nortriptyline in reducing OCD symptomatology, although this difference did not reach statistical significance (Thorén et al. 1980b). Ananth et al. (1981) reported that clomipramine, but not amitriptyline, produced significant amelioration of obsessions, depression, and anxiety. Volavka et al. (1985) reported clomipramine to be slightly more antiobsessional than imipramine at 12 weeks but probably not at 6 weeks of treatment, although differences in the baseline scores and in the symptom clusters between the two drug groups complicate the interpretation of these results. Clomipramine was found to be clearly superior to the monoamine oxidase inhibitor (MAOI) clorgyline in a crossover study (Insel et al. 1983).

Preliminary results from a recent placebo-controlled, parallel-designed double-blind multicenter study, which included more than 500 patients with OCD from 21 locations, confirm the superiority of clomipramine over placebo in the treatment of patients with OCD (DeVeaugh-Geiss 1988).

Taken together these studies suggest that clomipramine is more potent than placebo and that several excellent antidepressants, including clorgyline, nortriptyline, and amitriptyline, fail to reduce OCD symptoms. This apparent specificity of clomipramine for reducing OCD symptoms appears to be strikingly different from the roughly equivalent results with tricyclic drugs in the treatment of depression. However, because none of these studies has shown that clomipramine is clearly more effective than another tricyclic, we decided to perform the first study utilizing direct (versus parallel) comparison of

**Table 5-1.**  Double-blind studies of clomipramine in OCD

| Study | N | Design | Improvement in obsessive symptoms |
|---|---|---|---|
| **Comparing clomipramine to placebo** | | | |
| Thorén et al. 1980b | 35 (I) | parallel CMI (150 mg) vs. placebo | CMI > PLAC (5 weeks) |
| Marks et al. 1980 | 40 (I then O) | Parallel CMI (183 mg) vs. placebo | CMI > PLAC (4 weeks; self-rating only) |
| Montgomery 1980 | 14 | Crossover CMI (75 mg) vs. placebo | CMI > PLAC (4 weeks) |
| Flament et al. 1985 | 19 (I + O) | Crossover CMI (141 mg) vs. placebo | CMI > PLAC (5 + 5 weeks) childhood OCD |
| **Comparing clomipramine to other antidepressants** | | | |
| Thorén et al. 1980b | 35 (I) | Parallel CMI (150 mg) vs. NOR (150 mg) | CMI > PLAC; NOR not > PLAC (5 weeks) |
| Ananth et al. 1981 | 20 (I + O) | Parallel CMI (133 mg) vs. AMI (197 mg) | CMI, not AMI, improved symptoms |
| Insel et al. 1983 | 13 (I + O) | Crossover CMI (236 mg) vs. CLG (28 mg) | CMI > CLG (4 + 6 weeks); PLAC ineffective |
| Volavka et al. 1985 | 19 (O) then 16 (O) | Parallel CMI (275 mg) vs. IMI (265 mg) | CMI > IMI (12 weeks) |
| Zohar et al. 1987b | 10 (O) | Crossover CMI (235 mg) vs. DMI (290 mg) | CMI > DMI (6 weeks) |

*Note.*  N = number of patients, I = inpatients, O = outpatients. CMI = clomipramine; NOR = nortriptyline; PLAC = placebo; AMI = amitriptyline; CLG = clorgyline; IMI = imipramine; DMI = desipramine.

clomipramine to another tricyclic antidepressant. This study was designed as a double-blind, randomized crossover study of non-depressed patients with OCD, comparing clomipramine and desipramine. The crossover design, although statistically trouble-some, confers the advantage of using each subject as his or her own control and thus provides direct comparison of the effect of desipramine with the effect of clomipramine. Desipramine was chosen as the comparison drug because it predominantly inhibits norepinephrine synaptic uptake, in contrast to clomipramine, which has potent effects on serotonin (5-hydroxytryptamine [5-HT]) uptake (see below).

The distribution of maximum percent change from pretreatment baseline after treatment with clomipramine and desipramine in patients with OCD ($n = 10$ for both conditions) is shown in Figure 5-1. The results demonstrate that (1) clomipramine is significantly more potent than desipramine for reducing obsessive-compulsive (OC) symptoms, and that (2) desipramine, an excellent antidepressant, lacks significant antiobsessional effects. The differences in anti-obsessional effects between these two drugs are not caused by order effects (the drugs were given in a counterbalanced design) and cannot be related to intergroup diagnostic differences, as part of the analysis used each subject as his or her own control. Plasma levels for desipramine and clomipramine were roughly equivalent and within the therapeutic range for antidepressant effects.

It should be emphasized that the differences in improvement of OC ratings, although clearly significant, were not large. After 6 weeks of clomipramine treatment, only a 28% improvement in obsessional symptoms was noted. Other studies (Volavka et al. 1985; Zohar and Insel 1987b) have reported a relatively long time course for the development of maximal antiobsessional effects with clomipramine, with significant reductions at 4–6 weeks that did not become marked until 12 weeks. Eight of the patients in the current study (including the patient with 0% improvement) entered an open-label extended trial with clomipramine, leading to improvements of 40–90%. In each case, the symptoms remained to some extent, but were easier to resist and no longer interfered with functioning.

## CLOMIPRAMINE AND SEROTONIN

What then sets clomipramine apart pharmacologically from struc-turally related tricyclic antidepressants? Clomipramine differs from imipramine only by the chloride substituted at the number 3 carbon (Figure 5-2). As with many familiar compounds (such as am-phetamine and phenylalanine), this parachloro substitution confers a

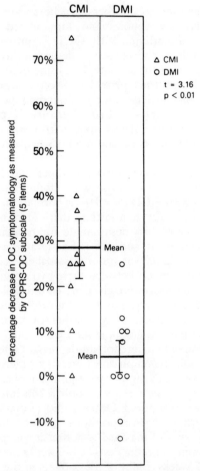

**Figure 5-1.** Distribution of maximum percent change from pretreatment baseline on clomipramine (CMI) and desipramine (DMI) in obsessive-compulsive (OC) patients ($n = 10$, for both trials), as measured by the Comprehensive Psychiatric Rating Scale — Obsessive-Compulsive (CPRS-OC) subscale developed by Thorén et al. (1980b) and modified by the present authors to score only five items specific for OC symptoms (i.e., rituals, compulsive thoughts, indecision, worrying over trifles, and lassitude). Reproduced from Zohar and Insel, *Biological Psychiatry* 22:667–687, 1987, with permission of the Elsevier Science Publishing Company. Copyright 1987, Elsevier Science Publishing Company.

number of different properties on the tricyclic ring. One of these properties is that clomipramine, compared to other tricyclics, is an extremely potent inhibitor of 5-HT reuptake (Ross and Renyi 1977). Moreover, even the main metabolite of clomipramine, desmethylclomipramine, which is present at about twice the concentration of clomipramine in the human plasma, maintains high affinity for the [3]H-labeled imipramine binding site, believed to be a 5-HT transporter (Paul et al. 1980). By way of comparison, the affinity of desipramine, the main metabolite of imipramine, for the [3H]imipramine binding site is an order of magnitude less than the affinity of desmethylclomipramine (Paul et al. 1980).

We compared the relationship between 5-HT and several tricyclic antidepressants in an in vivo model. The model used was the "5-HT syndrome" in rodents, which is considered a method of assessing direct effects on serotonergic transmission (Jacobs and Gelperin 1981; Wozniak 1984). In this model, clomipramine was found to be considerably more effective than imipramine, desipramine, amitriptyline, and nortriptyline in its potency for inducing the "5-HT syndrome" (Figure 5-3).

Several lines of evidence point to clomipramine's effect on 5-HT reuptake as most relevant to the drug's antiobsessional effects. Two studies have reported a significant correlation between plasma levels of clomipramine, but not of desmethylclomipramine (which has powerful inhibitory effects on the reuptake of both norepinephrine and 5-HT), and clinical improvement in OCD (Insel et al. 1983; Stern

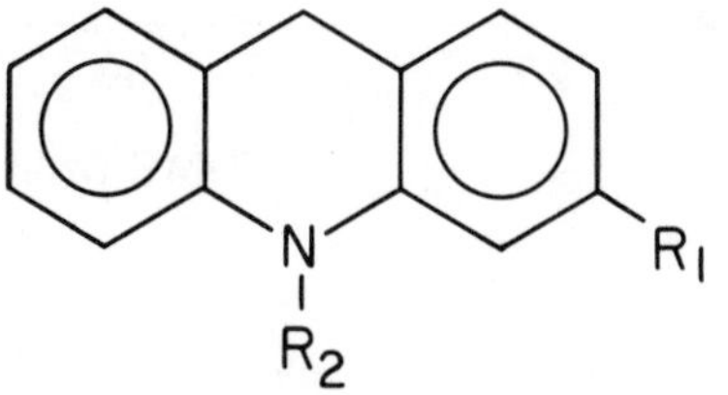

$R_1 = Cl$    $R_2 = CH_2-CH_2-CH_2-N(CH_3)_2$    Clomipramine

$R_1 = H$    $R_2 = CH_2-CH_2-CH_2-N(CH_3)_2$    Imipramine

$R_1 = Cl$    $R_2 = CH_2-CH_2-CH_2-N-CH_3$    Desmethylclomipramine

**Figure 5-2.** Chemical structure of clomipramine, imipramine, and desmethylclomipramine.

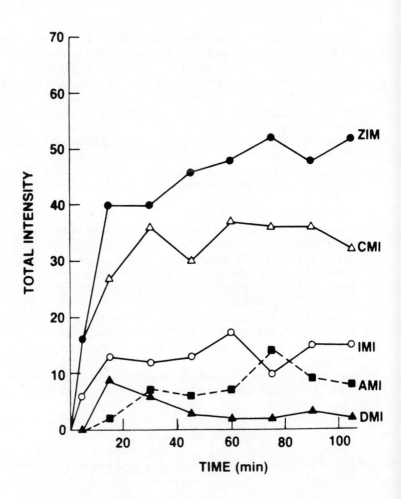

**Figure 5-3.** Comparison of the total intensity of the serotonergic syndrome in rats after administration of amine reuptake blockers. Five groups of 12 rats were treated with a nonselective MAOI, phenelzine (200 μmole/kg), and a second dose 16.5 hours later of 50 μmole/kg ip. After 90 minutes, each group was subsequently given a dose of 31.6 μmole/kg ip of an amine reuptake blocker (zimelidine [ZIM], clomipramine [CMI], imipramine [IMI], amitriptyline [AMI], and desipramine [DMI]). The total intensity of the serotonergic syndrome (Wozniak 1984) was graded every 15 minutes for 105 minutes.

et al. 1980; also see Flament et al. 1985; Thorén et al. 1980a). Thorén et al. (1980a) found a highly significant ($r = .75$) correlation between improvement in OCD symptoms during treatment with clomipramine and decrease in levels of the 5-HT metabolite 5-hydroxyindoleacetic acid (5-HIAA) in cerebrospinal fluid (CSF). Flament et al. (1985) similarly reported a highly significant correlation ($r = .78$) between clinical improvement and decrease in platelet 5-HT in children with OCD treated with clomipramine.

To investigate further the relationship between clomipramine's potent serotonergic effects and its clinical efficacy, Benkelfat et al. (1989) administered the 5-HT receptor antagonist metergoline to patients with OCD who had been receiving clomipramine on a long-term basis. These investigators found an average of 40% lessening in OC symptoms. In this crossover, placebo-controlled study, carried out under double-blind, random assignment conditions, the patients appeared to be modestly more anxious and obsessional on day 4 of metergoline administration compared to day 4 of placebo administration. Metergoline also lowered plasma prolactin concentration and thus provided evidence of physiologically significant 5-HT antagonism. However, it did not alter plasma clomipramine concentrations. This study appears to support further the hypothesis that the therapeutic effects of clomipramine in OCD are mediated via serotonergic mechanisms.

## CLINICAL IMPLICATIONS

When considered together, the above studies suggest the following: (1) The tricyclic antidepressant clomipramine reduces the symptoms of OCD, whereas a number of structurally related antidepressants, such as desipramine, imipramine, amitriptyline, and nortriptyline, do not. (2) The antiobsessional effects of clomipramine are mediated through the serotonergic system. One would expect, therefore, that other medications with potent 5-HT reuptake properties would also be effective in the treatment of OCD. Indeed, several recent studies have reported that nontricyclic selective 5-HT reuptake blockers, such as zimelidine (Fontaine and Chouinard 1986; Kahn et al. 1988; Prasad 1984; also see Insel et al. 1985), were effective antiobsessional agents; however, zimelidine has recently been declared unsafe for clinical use. Very recently, two selective 5-HT reuptake blockers, the bicyclic fluoxetine (Turner et al. 1985) and the unicyclic fluvoxamine (Goodman et al. 1989; Perse et al. 1987; Price et al. 1987), have been found to be effective in reducing OCD symptoms. Pending further studies, fluoxetine and fluvoxamine may be alternatives to clomipramine in the treatment of OCD.

Additionally, case reports have described augmentation of clomipramine's antiobsessional effect with the 5-HT precursor L-tryptophan (Rasmussen 1984), and with lithium (Eisenberg and Asnis 1985; Rasmussen 1984; Stern and Jenike 1983). In preclinical studies, lithium increased electrophysiological measures of serotonergic functioning in animals during chronic tricyclic treatment (Blier and de Montigny 1985).

As the data concerning the efficacy of clomipramine in OCD have expanded (Zohar and Insel 1987a), the importance of determining the necessary duration of treatment with clomipramine has become increasingly paramount. To examine the appropriate duration of treatment with clomipramine, a prospective placebo-controlled, double-blind discontinuation study was carried out (Pato et al. 1988). The study included 21 patients with OCD who manifested sustained improvement during 5 to 27 months of treatment with clomipramine. Of the 18 patients who completed the study, 16 patients experienced profound worsening of OCD symptoms after discontinuation of clomipramine. This worsening became statistically significant 4 weeks after discontinuation of the drug and appeared unrelated to clomipramine treatment duration prior to discontinuation, or to the type of OC symptoms originally presented. Reinstitution of clomipramine, in an open fashion, at the completion of the discontinuation study showed that the patients had returned to their pre-study levels of improvement. A possible implication of this study is that prolonged clomipramine administration is warranted for the majority of patients with OCD. However, further studies are needed to evaluate whether treatment for more than the period examined (mean: $11 \pm 5$ months) would allow for discontinuation without recurrence.

## OBSESSIVE-COMPULSIVE DISORDER AND SEROTONERGIC FUNCTION

Results from all of the above treatment studies are consistent with the hypothesis that the antiobsessional effects of pharmacological agents such as clomipramine, fluoxetine, and fluvoxamine are associated with these agents' potent serotonergic profile. However, whether patients have a psychobiological abnormality reflecting serotonergic functioning is an entirely different question. A way to examine this question is to compare drug-free patients with OCD with matched control subjects. This approach has been utilized in studies of $[^3H]$ imipramine binding in platelets and of 5-HIAA concentrations in CSF, and in observing behavioral and endocrine responses to pharmacologic challenges with selective serotonergic agonists.

High-affinity [$^3$H]imipramine binding sites in human platelets closely resemble the binding sites in human brains (Rehavi et al. 1984) and are closely associated with the presynaptic uptake site from 5-[$^3$H]HT (Paul et al. 1981). Thus, the high-affinity [$^3$H]imipramine binding may serve as a model for studying the central serotonergic neurons (Lingjaerde 1984). The maximum binding of the [$^3$H]imipramine binding sites was found to be significantly lower (50% lower, $P < .01$, Student's $t$ test) in 18 patients with OCD compared to 18 age- and sex-matched control subjects in one study (Weizman et al. 1986), although this was not found to be the case in a previous study (Insel et al. 1985).

The 5-HT metabolite 5-HIAA in CSF may serve as an indicator for central 5-HT turnover (for review, see Jimerson and Berrettini 1985). Significantly higher (30% higher, $P < .001$, Mann-Whitney U test) CSF 5-HIAA concentrations have been noted in a small cohort ($n = 8$) of patients with OCD compared to 5-HIAA values corrected for height and weight of 23 control subjects. In a previous, larger study (Thorén et al. 1980a), no significant increase in levels of CSF 5-HIAA was found, although there was a 20% increase in CSF 5-HIAA in patients with OCD compared to control subjects. The higher levels of CSF 5-HIAA in patients with OCD (vs. control subjects) are of interest because patients who are phenomenologically opposite to patients with OCD (impulsive and violent patients) have CSF 5-HIAA levels that are significantly lower than those in control subjects (Linnoila et al. 1983; van Praag 1983; Virkkunen et al. 1987).

To investigate further the possible role of 5-HT in the mediation of obsessional symptoms, we compared the behavioral, thermal, and endocrine effects following direct activation of the serotonergic system with a selective 5-HT receptor agonist in untreated patients with OCD and in healthy control subjects (Zohar et al. 1987b). Traditional drugs that affect serotonergic metabolism and action are not entirely specific for the serotonergic system (e.g., tryptophan or hydroxytryptophan) (Fernstrom and Wurtman 1974; Saavedra and Axelrod 1974) or are associated with severe adverse effects (e.g., *p*-chlorophenylalanine) (Engelman et al. 1974). Several selective 5-HT uptake inhibitors, such as zimelidine and citalopram, have been declared unsafe for clinical use, while some direct postsynaptic receptor agonists, such as dimethyltryptamine (Meltzer et al. 1982), are not useful because of their hallucinogenic properties. Therefore, we elected to study the novel 5-HT receptor agonist *m*-chlorophenylpiperazine (m-CPP), a synthetic, nonindole, aryl-substituted piperazine derivative that rapidly penetrates the blood-brain barrier (Fuller et al. 1980; Samanin et al. 1979). m-CPP is a metabolite of the antidepressant trazodone (Caccia et al. 1981, 1982) and thus might be expected to be safe and well tolerated in humans. In vitro studies with m-CPP show potent displacement of tritiated 5-HT from membrane homogenates (Fuller et al. 1981; Invernizzi et al. 1981). In vivo administration of m-CPP in preclinical biochemical studies of that drug led to decreased central 5-HT synthesis and turnover (Fuller et al. 1981; Samanin et al. 1980); this decrease was interpreted as a feedback effect following postsynaptic receptor stimulation. In animal studies, m-CPP produced typical changes in 5-HT-mediated be-haviors—for example, decreased food consumption (Mennini et al. 1980; Samanin et al. 1979) and decreased locomotion (Vetulani et al. 1982), neuroendocrine responses (increased prolactin, cortisol, and corticotropin levels [Aloi et al. 1984; Maj and Lewandowska 1980]), and physiologic reactions (hyperthermia [Aloi et al. 1984])—that are expected of a postsynaptic 5-HT receptor agonist. All these changes are reversed by the administration of the 5-HT receptor antagonist metergoline and are responsive to presumed alterations in 5-HT receptor sensitivity (Aloi et al. 1984; Maj and Lewandowska 1980; Samanin et al. 1979).

Using m-CPP as a serotonergic probe, we investigated, in a series of studies, whether (1) patients with OCD differed from healthy control subjects in their behavioral, endocrine, or thermal response to m-CPP, and (2) m-CPP would affect the symptoms of OCD. In addition, responses to m-CPP in the patients with OCD were com-pared to the responses in those patients on placebo and to those on metergoline, a 5-HT antagonist (Zohar et al. 1987b).

All studies consisted of orally administered, single-dose drug or placebo given under double-blind, random assignment conditions on separate days. The oral (versus intravenous) administration of m-CPP was chosen on the basis of concurrent studies in healthy subjects comparing levels of m-CPP (0.5 mg/kg) administered orally to those of m-CPP (0.1 mg/kg) administered intravenously (Murphy et al. 1989). The oral administration produced plateau-phase plasma m-CPP concentration and peak elevations of prolactin and cortisol that were essentially identical to the plateau-phase m-CPP concentra-tion and to the peak prolactin and cortisol elevations found after smaller doses of m-CPP (0.1 mg/kg) administered intravenously over 90 seconds. In contrast, striking differences in the subjective effects of m-CPP were observed in healthy control subjects. Intravenously administered m-CPP produced significant increases in anxiety, depressive affect, and derealization. After intravenous m-CPP ad-ministration, these subjects also experienced a sense of impaired cognitive capacity, while after larger doses of orally administered m-CPP, this cohort of healthy volunteers had smaller or negligible changes in all of these behavioral measures. Additionally, temperature elevations were greater after orally administered m-CPP, while somatic side effects were more pronounced after intravenously administered m-CPP. Because we wanted to minimize possible physical and mental side effects while still achieving reliable temperature and hormonal changes, we chose the oral administration.

The 0.5 mg/kg dose of m-CPP was chosen on the basis of concurrent studies in healthy control subjects with doses of m-CPP between 0.25 and 2.5 mg/kg. In contrast to the 0.5 mg/kg dose, which was generally well tolerated and which produced consistent increases in prolactin and cortisol levels and in temperature, doses of 1.0 mg/kg and greater produced transient nausea or vomiting in 6 of 13 subjects and also produced transient perceptual distortions, feelings of derealization, crying spells, anxiety, and/or dysphoria in several subjects (Mueller et al. 1985).

The effects of 0.5 mg/kg of m-CPP, administered orally to 12 drug-free patients with OCD and 14 control subjects, are shown in Figure 5-4. Relative to the healthy control subjects, the untreated patients with OCD became significantly more anxious, depressed, and dysphoric, and showed greater increases in altered self-reality (e.g., "out of touch," "mistrustful," "unusual thoughts," "unreal or strange"). The placebo response, as well as the physical effects in both groups, was very mild and did not significantly differ between the patients and the control subjects. This finding suggests that untreated patients with OCD are more responsive to the behavioral effects of m-CPP. However, it does not indicate whether m-CPP affects the symptoms of OCD. To test this effect, each patient completed, with the help of an experienced psychiatrist, two specific ratings aimed at assessing changes in obsessions and compulsions. The following scales were used: (1) a modified form of the Comprehensive Psychiatric Rating Scale—Obsessive-Compulsive (CPRS-OC) subscale developed by Thorén and coworkers (1980b); the modified CPRS-OC-5 includes only those five items (rated 0 to 3) that reflect OC symptoms (compulsive thoughts, rituals, indecision, lassitude, and concentration difficulties), and excludes three items (sadness, inner tension, and worrying over trifles) that are not specific to OCD; and (2) the NIMH Global scales, including those for OC symptoms, anxiety, and depression, rated from 1 to 15 (Murphy et al. 1982). The ratings all were done before and at 90, 180, and 210 minutes following drug or placebo administration.

Following administration of m-CPP, but not of placebo, untreated patients with OCD experienced a transient but marked exacerbation of their OC symptoms (Figure 5-5). The peak behavioral response was generally observed in the first 3 hours following m-CPP administration, and the duration of the worsening ranged from several hours (half of the patients) to 48 hours (one-fifth of the patients).

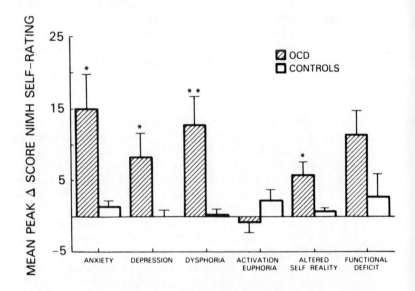

**Figure 5-4.**   Peak behavioral changes of patients with obsessive-compulsive disorder (OCD, shaded bars; $n = 12$) and control subjects (open bars; $n = 14$) after single-dose $m$-chlorophenylpiperazine administration (0.5 mg/kg) under randomized double-blind conditions. NIMH indicates National Institute of Mental Health. Anxiety: feels anxious, restless, worried, frightened; depression: feels sad, depressed, hopeless, worthless; dysphoria: feels irritable, angry, uncomfortable mentally; activation euphoria: feels elated, more talkative, especially energetic, has racing thoughts; altered self-reality: feels out of touch, mistrustful or suspicious, unreal or strange, has unusual thoughts, hears voices, sees things others don't; functional difficulty: has difficulty concentrating and functioning, feels slowed down. Asterisk indicates $P < .05$; double asterisk, $P < .01$. Student's $t$ test on peak behavioral changes, patients vs. control subjects. Reproduced from Zohar et al., *Archives of General Psychiatry* 44:946–951, 1987, with permission of the American Medical Association. Copyright 1987, American Medical Association.

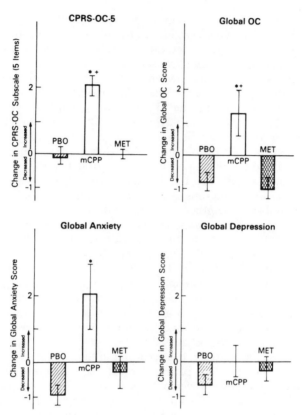

**Figure 5-5.** Peak changes ( ± SEM) from baseline on behavioral ratings after placebo (PBO; $n = 15$), *m*-chlorophenylpiperazine (m-CPP; $n = 12$), and metergoline (MET; $n = 8$) administered orally to obsessive-compulsive (OC) patients under double-blind conditions. CPRS-OC-5 indicates Comprehensive Psychiatric Rating Scale—Obsessive-Compulsive 5-item subscale; other scales are National Institute of Mental Health Global Scales. Asterisk indicates $P < .05$, Student's $t$ test of paired data comparing peak changes in behavioral measures after m-CPP administration with the peak changes in these measures after placebo in each subject. Plus sign indicates $P < .05$, Student's $t$ test comparing the peak changes in behavioral measures after m-CPP administration with the peak changes in these measures after metergoline. Reproduced from Zohar et al., *Archives of General Psychiatry* 44:946–951, 1987, with permission of the American Medical Association. Copyright 1987, American Medical Association.

Moreover, following administration of m-CPP, but not following administration of metergoline or placebo, five patients spontaneously described emergence of new obsessions or reoccurrence of obsessions that had not been present for several months.

Because oral administration of m-CPP (0.5 mg/kg) was also associated with anxiogenic effects and induction of panic attacks in patients with panic disorder (Kleine et al., submitted for publication) it is not clear whether m-CPP increases OC symptoms selectively or whether there may be a nonspecific exacerbation of diverse psychopathology by the drug. Certainly, m-CPP affected several symptoms other than obsessions in the patients with OCD, and thus one might assume that the drug increases anxiety or dysphoria, and increases OC symptoms as a secondary phenomenon. This assumption would be strengthened if patients with OCD were hypersensitive to other anxiogenic compounds. However, several recent studies with anxiogenic agents such as lactate (Gorman et al. 1985), yohimbine (Rasmussen et al. 1987), and caffeine (Zohar et al. 1987a) found that these agents failed to increase either anxiety or OC symptoms in patients with OCD. Therefore, it seems that patients with OCD are not especially prone to respond to provocative agents that exacerbate anxiety disorders. Moreover, m-CPP is, so far, the only agent associated with exacerbation of OC symptoms (Zohar et al. 1987b).

In contrast to the behavioral hypersensitivity, equivalent (although significant) increases in prolactin and in temperature were observed in both groups. Peak plasma cortisol concentrations after m-CPP administration were lower in the patients than in the control subjects. The patients also tended to have lower baseline cortisol values than the control subjects. The lack of a differentially increased secretion of "stress hormones," such as prolactin and cortisol, in the presence of an apparently very stressful experience in patients with OCD is a puzzling phenomenon. One possible interpretation for this response is that the effects of m-CPP on behavior are mediated by a different mechanism than the one that mediates the effect of m-CPP on endocrine and thermal regulation. The lack of difference in baseline prolactin or cortisol levels and temperature between patients and control subjects is consistent with this interpretation, which implies that in patients with OCD the entire serotonergic system is not affected, but only those elements that mediate the OC symptoms. Another possible interpretation is that the effects of m-CPP bypass the normal stress control mechanisms that regulate the increased secretion of prolactin and cortisol. A third interpretation is that the patients with OCD are hormonally hyporesponsive. This interpretation was hinted at by the reduced cortisol responses to m-CPP in patients with OCD as compared to patients with panic disorder, who exhibit increased cortisol responses to 0.25 mg/kg m-CPP administered orally (Kahn et al. 1988).

## SEROTONIN HYPOTHESIS OF OBSESSIVE-COMPULSIVE DISORDER: NEW FORMULATION

One might wonder why a 5-HT agonist like m-CPP is associated with acute exacerbation of OC symptoms, while agents such as clomipramine, fluvoxamine, and fluoxetine, which are believed to increase serotonergic function indirectly, decrease OC symptoms. This apparent paradox might be explained if, as has been reported in animal studies, long-term treatment with these medications leads to adaptive downregulation of 5-HT receptors (Blackshear and Sanders-Bush 1982; Fuxe et al. 1983; Olpe et al. 1984).

To investigate possible changes in brain serotonergic responsivity, m-CPP and placebo were given under double-blind conditions to nine patients with OCD before and after treatment with clomipramine (Zohar et al. 1988). Unlike our previous observations of a marked transient increase in obsessional symptoms and anxiety following administration of m-CPP in untreated patients, readministration of m-CPP after 4 months of treatment with clomipramine did not significantly increase obsessional symptoms and anxiety. Similarly, the hyperthermic effect of m-CPP observed before treatment was eliminated after treatment with clomipramine.

The ability of clomipramine to reduce OC symptoms in patients with OCD coupled with the development of serotonergic hyporesponsivity during clomipramine treatment raises the possibility that the therapeutic effects of clomipramine in patients with OCD may be associated with the development of serotonergic subsensitivity. Two previous studies are consistent with this hypothesis. Thorén et al. (1980a) reported that the clinical response of patients with OCD treated with clomipramine was positively correlated ($r = .75$) with the decrease in CSF concentration of 5-HIAA, the main 5-HT metabolite, but not with changes in 3-methoxy-4-hydroxyphenylglycol (MHPG) or homovanillic acid (HVA). Flament et al. (1985) noted that clinical improvement as measured in children with OCD by using several rating scales was significantly correlated ($r = .62$ to .78) with decreases in platelet 5-HT concentration. In the present study, probably due to the small number of patients, we did not find a significant correlation between the therapeutic response to clomipramine and the magnitude of the subsensitivity to m-CPP that developed in individual patients, although as a group the patients had beneficial clinical responses and reduced serotonergic responsivity during clomipramine treatment.

In the classical "OCD serotonergic hypothesis," OCD was linked to a 5-HT deficiency state, because patients had favorable responses to the 5-HT reuptake blocker clomipramine (Yaryura-Tobias et al. 1978). Additional support for the 5-HT deficiency hypothesis came from observations that whole-blood 5-HT levels were reduced in patients with OCD (Yaryura-Tobias et al. 1978), and from uncontrolled trials of the 5-HT precursor L-tryptophan, which was reported to reduce obsessional symptoms (Yaryura-Tobias and Bhagavan 1977) and to augment the antiobsessional effect of clomipramine (Rasmussen 1984). Based on this hypothesis, one might expect that administration of a 5-HT agonist (like m-CPP) to patients with OCD would reduce OC symptoms, at least briefly. However, m-CPP has been observed to increase rather than decrease OC symptoms in these patients (Zohar et al. 1987b). Moreover, beneficial clomipramine treatment in patients with OCD was associated with decreased serotonergic responsivity in patients with OCD compared with the same patients' responses to the same serotonergic challenge (0.5 mg/kg m-CPP) before treatment (Zohar et al. 1988).

These findings are all consistent with the hypothesis that increased serotonergic responsivity, rather than 5-HT deficiency, is associated with the psychopathological characteristics of OCD. This formulation of the "serotonergic hypothesis" for OCD is also consistent with the relatively long period of time needed for selective treatment with 5-HT reuptake blockers to become effective (Insel and Zohar 1987).

However, if hyper-responsivity of the serotonergic system is associated with the psychopathological characteristics of OCD, we might also expect to see worsening of OCD symptoms during the initial phase of treatment with 5-HT reuptake blockers, because initially these medications increase 5-HT content in the synaptic cleft. A recent reanalysis of daily ratings of patients from an earlier inpatient study (Insel et al. 1983) provides preliminary support for the notion that many patients with OCD experience exacerbation of their symptoms during the first 3 to 5 days of clomipramine treatment (J. Zohar, T. R. Insel, and D. L. Murphy, unpublished findings).

## SUMMARY AND FUTURE DIRECTIONS

The "5-HT hypothesis" for OCD was put forward primarily on the basis that clomipramine, a potent 5-HT reuptake blocker, was found to be more effective than other structurally related tricyclic antidepressants in reducing OCD symptoms. Recently, other nontricyclic, potent 5-HT reuptake blockers such as fluvoxamine and fluoxetine also have been reported to have antiobsessional properties. m-CPP, a novel serotonergic agonist, administered orally to untreated patients with OCD produced markedly increased behavioral responses relative to control subjects, with a transient, but marked exacerbation of their OC symptoms. Clomipramine treatment was associated with decreased serotonergic responsivity in these patients. These findings are consistent with increased, rather than decreased, serotonergic responsivity in patients with OCD. While 5-HT seems to play a critical role in the psychobiology of OCD, the focus on a single neurotransmitter (or a single receptor) is undoubtedly an over-simplification of this intriguing disorder. The task ahead is to understand the relationship of a single variable, such as 5-HT, to the multitude of neurobiological, anatomical, and psychological variables that contribute to this disorder.

# REFERENCES

Akhtar S, Wig NH, Verma VK, et al: A phenomenological analysis of symptoms in obsessive compulsive neurosis. Br J Psychiatry 127:342–348, 1975

Aloi JA, Insel TR, Mueller EA, et al: Neuroendocrine and behavioral effects of *m*-chlorophenylpiperazine administration in rhesus monkeys. Life Sci 34:1325–1331, 1984

Ananth J, Solyom L, Bryntwick S, et al: Clorimipramine therapy for obsessive compulsive neurosis. Am J Psychiatry 136:700–701, 1979

Ananth J, Pecknad JC, Vandersteen N, et al: Double blind comparative study of clomipramine and amitryptyline in obsessive neurosis. Prog Neuro-psychopharmacol Biol Psychiatry 5:257–269, 1981

Benkelfat C, Murphy DL, Zohar J, et al: Clomipramine in obsessive-compulsive disorder: further evidence for a serotonergic mechanism of action. Arch Gen Psychiatry 46:23–28, 1989

Blackshear MA, Sanders-Bush E: Serotonin receptor sensitivity after acute and chronic treatment with mianserin. J Pharmacol Exp Ther 221:303–308, 1982

Blier P, de Montigny C: Short-term lithium administration enhances serotonergic neurotransmission: electrophysiological evidence in the rat CNS. Eur J Pharmacol 113:69–79, 1985

Caccia S, Ballabio M, Samanin RJ: *m*-Chlorophenylpiperazine, a central 5-hydroxytryptamine agonist, is a metabolite of trazodone. J Pharm Pharmacol 33:477–478, 1981

Caccia S, Fong MH, Garattini S, et al: Plasma concentration of trazodone and 1-(3-chlorophenyl)piperazine in man after a single oral dose of trazodone. J Pharm Pharmacol 34:605–606, 1982

DeVeaugh-Geiss J: A multicenter trial of Anafranil in obsessive compulsive disorder. Abstract of paper presented at the 141st annual meeting of the American Psychiatric Association, Montreal, May 1988, pp 259–260

Eisenberg J, Asnis C: Lithium as an adjunct treatment in obsessive-compulsive disorder. Am J Psychiatry 142:663, 1985

Engelman K, Lovenberg W, Sjoerdsma A: Inhibition of serotonin by parachlorophenylalanine in patients with carcinoid syndrome. N Engl J Med 277:1103–1108, 1974

Fernstrom S, Wurtman RJ: Nutrition and the brain. Sci Am 230:85–96, 1974

Flament MF, Rapoport JL, Berg CJ, et al: Clomipramine treatment of childhood obsessive-compulsive disorder: a double blind controlled study. Arch Gen Psychiatry 42:977–983, 1985

Flament M, Whitaker A, Rapoport JL, et al: An epidemiological study of obsessive-compulsive disorder in adolescence, in Obsessive-Compulsive Disorder in Children and Adolescents. Edited by Rapoport JL. Washington, DC, American Psychiatric Press, 1989, pp 253–267

Fontaine R, Chouinard G: An open clinical trial of fluoxetine in the treatment of obsessive-compulsive disorder. J Clin Psychopharmacol 6:98–100, 1986

Fuller RW, Mason NR, Molloy BP: Structural relationships in the inhibition of $^3$H-serotonin binding to rat membranes in vitro by 1-phenyl-piperazines. Biochem Pharmacol 29:833–835, 1980

Fuller RW, Snoddy HD, Mason NR, et al: Disposition and pharmacological effects of *m*-chlorophenylpiperazine in rats. Neuropharmacology 20:155–162, 1981

Fuxe K, Ogren SO, Agnati NF, et al: Chronic antidepressant treatment and central 5-HT synapses. Neuropharmacology 22:389–400, 1983

Goodman WK, Price LH, Rasmussen SA, et al: Efficacy of fluvoxamine in obsessive-compulsive disorder: a double-blind comparison with placebo. Arch Gen Psychiatry 46:36–44, 1989

Gorman JM, Liebowitz MR, Fyer AJ, et al: Lactate infusions in obsessive-compulsive disorder. Am J Psychiatry 142:864–866, 1985

Insel TR: Obsessive-compulsive disorder: the clinical picture, in New Findings in Obsessive-Compulsive Disorder. Edited by Insel TR. Washington, DC, American Psychiatric Press, 1984, pp 1–22

Insel TR, Zohar J: Psychopharmacologic approaches to obsessive compulsive disorder, in Psychopharmacology: The Third Generation of Progress. Edited by Meltzer HY. New York, Raven Press, 1987, pp 1205–1210

Insel TR, Murphy DL, Cohen RM, et al: Obsessive-compulsive disorder: a double blind trial of clomipramine and clorgyline. Arch Gen Psychiatry 40:605–612, 1983

Insel TR, Mueller EA, Alterman I, et al: Obsessive-compulsive disorder and serotonin: is there a connection? Biol Psychiatry 20:1174–1185, 1985

Invernizzi R, Cotecchia S, DeBlasi A, et al: Effects of *m*-chlorophenyl-piperazine on receptor binding and brain metabolism of monoamines in rats. Neurochem Int 3:239–244, 1981

Jacobs BL, Gelperin A: Serotonin Neurotransmission and Behavior. Cambridge, MA, MIT Press, 1981

Jimerson DC, Berrettini W: Cerebrospinal fluid metabolite studies in depression: research update, in Pathochemical Markers in Major Psychoses. Edited by Beckmann H, Riederer P. Berlin, Springer-Verlag, 1985, pp 129–143

Kahn RS, Asnis JM, Wetzler S, et al: Neuroendocrine evidence for serotonin receptor hypersensitivity in panic disorder. Psychopharmacology (Berlin) 96:360–364, 1988

Karno M, Golding JM, Sorenson SB, et al: The epidemiology of obsessive-compulsive disorder in five US communities. Arch Gen Psychiatry 45:1094–1099, 1988

Kleine E, Zohar J, Geraci MF, et al: Anxiogenic effects of m-CPP in patients with panic disorder: comparison to caffeine's anxiogenic effects (submitted for publication)

Lingjaerde O: Blood platelets as a model system for studying the biochemistry of depression, in Frontiers in Biochemical and Pharmacological Research in Depression: Advances in Biochemical Psychopharmacology. Edited by Usdin E, Åsberg M, Bertilsson L, et al. New York, Raven Press, 1984, pp 173–182

Linnoila M, Virkkunen M, Scheinin M, et al: Low cerebrospinal fluid 5-HIAA concentration differentiates impulsive from non-impulsive violent behavior. Life Sci 33:2609–2614, 1983

Maj J, Lewandowska A: Central serotoninmimetic action of phenyl-piperazine. Pol J Pharmacol Pharm 32:495–504, 1980

Marks IM, Stern R, Mawson D, et al: Clomipramine and exposure for obsessive-compulsive rituals. Br J Psychiatry 136:1–25, 1980

Meltzer H, Wiita B, Tricou BJ, et al: Effect of serotonin precursors and serotonin agonists on plasma hormone levels, in Serotonin in Biological Psychiatry. Edited by Ho BT, Schoolar JC, Usdin E. New York, Raven Press, 1982, pp 117–136

Mennini T, Poggasi E, Caccia S, et al: The endocrinology of stress. Aviat Space Environ Med 56:642–650, 1980

Montgomery SA: Clomipramine in obsessional neurosis: a placebo-controlled trial. Psychol Med 1:189–192, 1980

Mueller EA, Murphy DL, Sunderland T: Neuroendocrine effects of m-chlorophenylpiperazine, a serotonin agonist in humans. J Clin Endocrinol Metab 61:1–6, 1985

Murphy DL, Pickar D, Alterman I: Methods for the quantitative assessment of depressive and manic behavior, in The Behavior of Psychiatric Patients. Edited by Burdock EI, Sudilovsky A, Gershon S. New York, Marcel Dekker, 1982, pp 355–391

Murphy DL, Mueller EA, Hill JL, et al: Comparative anxiogenic, neuroendocrine, and other physiologic effects of m-chlorophenylpiperazine given intravenously or orally to healthy volunteers. Psychopharmacology (Berlin) 98:275–282, 1989

Olpe HR, Schellenberg A, Jones RSG: The sensitivity of hippocampal pyramidal neurons to serotonin in vitro: effect of prolonged treatment with clorgyline or clomipramine. J Neural Transm 60:265–271, 1984

Pato MT, Zohar-Kadouch RC, Zohar J, et al: Return of symptoms after discontinuation of clomipramine in patients with obsessive-compulsive disorder. Am J Psychiatry 145:1521–1525, 1988

Paul SN, Rehavi M, Hulihan B, et al: A rapid and sensitive radioreceptor assay for tertiary amine tricyclic antidepressants. Commun Psychopharmacol 4:487–494, 1980

Paul SN, Rehavi N, Rice KC, et al: Does high affinity [3]H-imipramine binding label serotonin reuptake sites in brain and platelet? Life Sci 28:2753–2760, 1981

Perse TI, Greist JH, Jefferson JW, et al: Fluvoxamine treatment of obsessive compulsive disorder. Am J Psychiatry 144:1059–1061, 1987

Prasad A: Obsessive-compulsive disorders and trazodone. Am J Psychiatry 141:612–613, 1984

Price LH, Goodman WK, Charney DS, et al: Treatment of severe obsessive compulsive disorder with fluvoxamine. Am J Psychiatry 144:1059–1061, 1987

Rachman SJ: Primary obsessional slowness. Behav Res Ther 11:463–471, 1974

Rachman SJ, Hodgson RJ: Conventional treatment and prognosis, in Obsessions and Compulsions. Englewood Cliffs, NJ, Prentice-Hall, 1980, pp 97–105

Rack PH: Clinical experience in the treatment of obsessional states. J Int Med Res 5:81–96, 1977

Rasmussen SA: Lithium and tryptophan augmentation in clomipramine-resistant obsessive-compulsive disorder. Am J Psychiatry 141:1283–1285, 1984

Rasmussen SA, Tsuang MT: Epidemiology of obsessive-compulsive disorder: a review. J Clin Psychiatry 45:450–457, 1984

Rasmussen SA, Goodman WK, Woods SW, et al: Effects of yohimbine in obsessive-compulsive disorder. Psychopharmacology (Berlin) 93:308–313, 1987

Rehavi M, Paul SM, Skolnick P: High affinity binding sites for tricyclic antidepressant in brain and platelets, in Brain Receptor Methodologies, Part B. Edited by Marangos PJ, Campbell IC, Cohen RM. Orlando, FL, Academic Press, 1984, pp 81–89

Renynghe de Voxurie GE: Anafranil (G 34586) in obsessive neurosis. Acta Neurol Belg 68:787–792, 1968

Robins LN, Heltzer VE, Weissman MM, et al: Lifetime prevalence of psychiatric disorder in three communities. Arch Gen Psychiatry 41:949–967, 1984

Ross SB, Renyi AL: Inhibition of the neuronal uptake of 5-hydroxytryptamine and noradrenaline in rat brain by (Z) and (E)-3-(4-bromophenyl)-N, N–dimethyl-3(3-pyridil) allylamines and their secondary analogues. Neuropharmacology 16:57–63, 1977

Saavedra J, Axelrod J: Brain tryptamine and the effect of drugs. Adv Biochem Psychopharmacol 10:135–139, 1974

Salzman L, Thaler FH: Obsessive-compulsive disorder: a review of the literature. Am J Psychiatry 138:286–296, 1981

Samanin R, Mennini T, Ferraris A: *m*-Chlorophenylpiperazine: a central serotonin agonist causing powerful anorexia in rats. Naunyn Schmiedebergs Arch Pharmacol 308:159–163, 1979

Samanin R, Caccia S, Bendotti C, et al: Further studies on the mechanism of the serotonin dependent anorexia in rats. Psychopharmacology (Berlin) 68:99–104, 1980

Stern TA, Jenike MA: Treatment of obsessive-compulsive disorder with lithium carbonate. Psychosomatics 24:674–683, 1983

Stern RS, Marks IM, Wright J, et al: Clomipramine: plasma levels, side effects and outcome in obsessive compulsive neurosis. Postgrad Med J 56:134–139, 1980

Thorén P, Åsberg M, Bertilsson L, et al: Clomipramine treatment of obsessive-compulsive disorder. II. Biochemical aspects. Arch Gen Psychiatry 37:1289–1294, 1980a

Thorén P, Åsberg M, Cronholm B, et al: Clomipramine treatment of obsessive-compulsive disorder. I. A controlled clinical trial. Arch Gen Psychiatry 37:1281–1285, 1980b

Turner SM, Jacob RG, Beidel DC, et al: Fluoxetine treatment of obsessive-compulsive disorder. J Clin Psychopharmacol 5:207–212, 1985

van Praag HM: CSF 5-HIAA and suicide in non-depressed schizophrenics. Lancet 2:977–978, 1983

Vetulani J, Sansone M, Bednarczyk B, et al: Different effects of 3-chlorophenylpiperazine on locomotor activity and acquisition of conditioned avoidance response in different strains of mice. Naunyn Schmiedebergs Arch Pharmacol 392:271–274, 1982

Virkkunen M, Nuutila A, Goodwin F, et al: Cerebrospinal fluid monoamine metabolite levels in male arsonists. Arch Gen Psychiatry 44:241–247, 1987

Volavka J, Neziroglu F, Yaryaru-Tobias JA: Clomipramine and imipramine in obsessive-compulsive disorder. Psychiatry Res 14:83–91, 1985

Weizman A, Carmi M, Hermesh H, et al: High affinity imipramine binding and serotonin uptake in platelets of eight adolescent and ten adult obsessive-compulsive patients. Am J Psychiatry 143:335–339, 1986

Welner A, Reich T, Robins E, et al: Obsessive-compulsive neurosis. Compr Psychiatry 17:527–539, 1976

Woodruff R, Pitts F: Monozygotic twins with obsessional illness. Am J Psychiatry 120:1075–1080, 1969

Wozniak KM: Interaction between inhibitors of monoamine oxidase and amine reuptake in rats. Unpublished doctoral dissertation, Department of Pharmacology, Institute of Psychiatry, University of London, London, 1984

Yaryura-Tobias JA, Bhagavan HN: L-Tryptophan in obsessive-compulsive disorders. Am J Psychiatry 234:1298–1299, 1977

Yaryura-Tobias JA, Neziroglu F: The action of clomipramine in obsessive compulsive neurosis: a pilot study. Curr Ther Res 17:111–116, 1975

Yaryura-Tobias JA, Bebirian RJ, Neziroglu F, et al: Obsessive-compulsive disorder as a serotonin defect. Res Commun Psychol Psychiatry Behav 2:279–286, 1978

Zohar J, Insel TR: Drug treatment of obsessive-compulsive disorder. J Affective Disord 13:193–202, 1987a

Zohar J, Insel TR: Obsessive-compulsive disorder: psychobiological approaches to diagnosis, treatment and pathophysiology. Biol Psychiatry 22:667–687, 1987b

Zohar J, Klein EM, Mueller EA, et al: 5HT, obsessive compulsive disorder and anxiety. Abstract of paper presented at the 140th annual meeting of the American Psychiatric Association, Chicago, IL, May 1987a, p 175

Zohar J, Mueller EA, Insel TR, et al: Serotonergic responsivity in obsessive-compulsive disorder: comparison of patients and healthy controls. Arch Gen Psychiatry 44:946–951, 1987b

Zohar J, Insel TR, Zohar-Kadouch RC, et al: Serotonergic responsivity in obsessive-compulsive disorder: effects of chronic clomipramine treatment. Arch Gen Psychiatry 45:167–172, 1988

# Chapter 6

## Hypotheses Relating Serotonergic Dysfunction to the Etiology and Treatment of Panic and Generalized Anxiety Disorders

Dennis S. Charney, M.D.
Scott W. Woods, M.D.
John H. Krystal, M.D.
Linda M. Nagy, M.D.
George R. Heninger, M.D.

# Chapter 6

## Hypotheses Relating Serotonergic Dysfunction to the Etiology and Treatment of Panic and Generalized Anxiety Disorders

In recent years there has been considerable research designed to increase the understanding of the neurobiologic etiology of anxiety disorders and the mechanism of action of antianxiety drugs. Although most of the clinical research has focused on establishing that a dysfunction of noradrenergic neuronal systems is involved in the origin and treatment of these disorders, there is preclinical evidence suggesting that alterations in serotonin (5-hydroxytryptamine [5-HT]) function may relate to the development of certain types of anxiety and to the therapeutic effectiveness of specific antianxiety drugs. In addition, there are several recent clinical reports that have assessed 5-HT function in anxiety disorder patients before and during antianxiety drug treatment. In this chapter we will review preclinical and clinical investigations that impact on the hypothesis that 5-HT function is important in the genesis and treatment of anxiety disorders.

## PRECLINICAL INVESTIGATIONS

### Neuroanatomy of the Brain Serotonergic System

Neuroanatomical studies of the serotonergic system indicate a distribution consistent with an important role for this system in behavioral regulation. Numerous investigations using retrograde and anterograde fiber-tracing techniques or immunohistochemical methods have shown that the cell bodies of serotonergic neurons are located in the midbrain raphe nuclei. The neurons in the dorsal and median raphe nuclei give rise to major ascending projections to the cerebral cortex, basal ganglion, and limbic areas (Figure 6-1), while neurons located in the other raphe nuclei innervate mainly the spinal cord and the cerebellum (Parent et al. 1981; Steinbusch 1981).

## Serotonin Receptor Subtypes

There is a convincing body of evidence that several types of 5-HT receptors exist in the mammalian brain. Three of these types of receptors that have been characterized are designated $5\text{-HT}_1$, $5\text{-HT}_2$, and $5\text{-HT}_3$ sites. The $5\text{-HT}_1$ sites are currently further subdivided into at least four distinct subsets that differ in their regional distribution and function and are termed $5\text{-HT}_{1A}$, $5\text{-HT}_{1B}$, $5\text{-HT}_{1C}$, and $5\text{-HT}_{1D}$ sites (Peroutka 1988).

The anatomical distribution of $5\text{-HT}_1$ receptors in human postmortem brain tissue has been studied by quantitative light-micro-

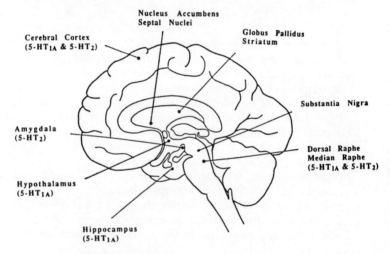

**Figure 6-1.**   Approximate anatomical representation of the main ascending serotonergic projections of the dorsal and median raphe nuclei. Many of the brain areas receiving serotonergic innervation have been hypothesized to be involved in the development of anxiety or fear. Brain areas containing high densities of $5\text{-HT}_{1A}$ and $5\text{-HT}_2$ receptors are indicated by parenthetical notation. Areas containing high $5\text{-HT}_{1A}$ densities may be sites of action for $5\text{-HT}_{1A}$ agonists such as buspirone, and areas with dense $5\text{-HT}_2$ localization may be sites of action for $5\text{-HT}_2$ antagonists such as ritanserin.

scopic autoradiography (Pazos et al. 1987a). Confirming previous findings, $5\text{-HT}_{1A}$ and $5\text{-HT}_{1C}$ receptors have been found in human brain, while sites with pharmacological characteristics of $5\text{-HT}_{1B}$ binding sites have not been definitively identified. The distribution of $5\text{-HT}_{1A}$ and $5\text{-HT}_{1C}$ receptor subtypes throughout the human brain is heterogeneous. High or very high densities of $5\text{-HT}_{1A}$ receptors are located in the hippocampus, raphe nuclei, layers I and II of the cortex, and nuclei of the thalamus and the amygdala. Intermediate levels of these receptors are found in the claustrum, posterior hypothalamus, mesencephalic and pontine central gray matter, and substantia gelatinosa of the cervical spinal cord. High densities of $5\text{-HT}_{1C}$ receptors are present in the choroid plexus, substantia nigra, globus pallidus, and ventral medial hypothalamus, while low or very low amounts of this receptor subtype are found in most of the other human brain areas studied. The anatomical distribution of $5\text{-HT}_{1D}$ receptors has not been well characterized.

Analysis of the distribution of $5\text{-HT}_1$ receptors in relation to concentrations of endogenous 5-HT and the presence of serotonergic terminals in the primate brain indicates that while the substantia nigra and the raphe nuclei are very enriched in 5-HT and $5\text{-HT}_1$ receptors, other areas such as the hippocampus and neocortex that have high densities of $5\text{-HT}_{1A}$ receptors are poor in endogenous 5-HT. Most of the brain areas and serotonergic nuclei containing $5\text{-HT}_1$ receptors have been shown to receive nerve endings from the serotonergic cells of the raphe nuclei of the brain stem. These regions include the hippocampus, basal ganglia, substantia nigra, choroid plexus, neocortex, and hypothalamus, among others. Therefore, a good correlation appears to exist between serotonergic innervation and the distribution of $5\text{-HT}_1$ receptors in the human brain (Pazos et al. 1987a).

The anatomical distribution of $5\text{-HT}_2$ receptors in the human brain has been studied by light-microscopic autoradiography using ketanserin as a ligand (Pazos et al. 1987b). Very high concentrations have been localized in layers III and V in several cortical areas, including the frontal, parietal, temporal, and occipital lobes, as well as the corpus mamillare of the hypothalamus. The claustrum, the nucleus lateralis of the amygdala, and some cortical layers also have high densities of $5\text{-HT}_2$ receptors. Intermediate concentrations of these receptors are found in the hippocampus, caudate, putamen, and nucleus accumbens, and in some nuclei of the amygdala, among other structures. Brain areas such as the thalamus, brain stem, cerebellum, and spinal cord contain, in general, only low to very low densities of $5\text{-HT}_2$ receptors.

The distribution of $5\text{-HT}_2$ receptors shows only a weak correlation with endogenous 5-HT concentration. Areas such as the substantia nigra, raphe nuclei, and thalamus, which have the highest concentration of 5-HT, contain only intermediate to low densities of 5-HT$_2$ receptors. Moreover, the cerebral cortex, which is the area most enriched in 5-HT$_2$ receptors, is poor in 5-HT. Studies of the distribution of serotonergic terminals indicate that the brain areas having high densities of 5-HT$_2$ receptors, such as the cerebral cortex, corpus mamillare of the hypothalamus, striatum, midline thalamic nuclei, hippocampal complex, amygdala, and ventral horn of the spinal cord, contain nerve terminals arising from the various groups of serotonergic cells in the brain stem, mainly, the nuclei raphe dorsalis, raphe centralis, raphe pontis, and raphe magnus. On the other hand, many brain areas enriched in 5-HT terminals, including the septum, substantia nigra, cerebellum, and locus coeruleus, are poor in 5-HT$_2$ receptors (Pazos et al. 1987b).

The anatomical distribution of 5-HT$_3$ receptors in human brain has been determined in several studies. Very high levels of 5-HT$_3$ receptors have been found in discrete nuclei of the lower brain stem, including the area postema, nuclei of the solitary tract, vagus nerve, and the spinal trigeminal nucleus. Considerable binding was also found in the substantia gelatinosa at all levels of the spinal cord (Waeber et al. 1989). It has also been demonstrated that [$^3$H]zacopride labels high-affinity 5-HT$_3$ receptor binding sites in the human hippocampus and amygdala (Barnes et al. 1989).

## Neuroanatomical Substrates of Anxiety or Fear: Relation to the Serotonin Hypothesis of Anxiety

A comprehensive discussion of the neural substrates of anxiety or fear is beyond the scope of this chapter. However, briefly outlined below is neuroanatomical, neurophysiological, and neurochemical evidence supporting a role for several brain areas in the development of anxiety or fear (see LeDoux 1987 for a review of this topic; see also Table 6-1).

From a neuroanatomical perspective, the amygdala is particularly well suited to be an important factor in the mediation of anxiety or fear. The amygdala receives afferents from cortical and thalamic exteroceptive systems, as well as from subcortical visceral afferent pathways. Projections from the amygdala go to autonomic pathways (lateral hypothalamus), neurohumoral pathways (paraventricular-supraoptic region, medial basal hypothalamus), and skeletal motor systems (ventral tegmental area to the caudate, putamen, and substantia nigra) (Jones and Powell 1970; Mesulam et al. 1977; Swanson 1983). The amygdala may, therefore, serve as a general homeostatic

**Table 6-1.**   Hypothesized neuroanatomical substrates of human anxiety

### Amygdala

Receives afferents from cortical and thalamic exteroceptive systems and cortical pathways.

Projects to autonomic pathways, neurohumeral pathways, and skeletomotor systems.

Amygdalectomy results in:
  No response to threatening stimuli in monkeys
  Increased punished responding in rats
  Blockade of conditioned fear in rats

### Locus coeruleus

Pharmacologic agents with anxiogenic properties increase, and anxiolytics decrease, locus coeruleus (LC) activity.

Uncontrollable stress increases LC firing rate in rats.

Electrical or pharmacological stimulation of the LC induces fear in monkeys.

LC firing rates in freely moving cats correlate with experimentally induced fear.

### Thalamus

Classically conditioned fear involves the relay of sensory signals to limbic forebrain from thalamus.

Fear responses to acoustic stimuli are disrupted by thalamic lesions.

### Hippocampus

Septohippocampal system has connections with limbic structures and cortical sensory areas.

Hippocampal lesions, like antianxiety drugs, increase punished responding.

### Neocortical areas

Neocortical sensory areas provide inputs to amygdala and hippocampus.

These areas probably play a role in the processing of complex stimulus information (e.g., stimuli involving threat).

function evaluating the significance of input from exteroceptive and interoceptive sources and initiating behavioral and visceral responses accordingly. Amygdalectomy has been shown to reduce the response to threatening stimuli in monkeys, to increase punished responding in rats, and to block the conditioned fear induced by the potentiated startle response (Blanchard and Blanchard 1972; Cohen 1980; Downer 1961; Hitchcock and Davis 1986; Iwata et al. 1986; Kapp et al. 1979, 1984). The presence of 5-HT1A, 5-HT3, benzodiazepine, noradrenergic, and opiate receptors on the amygdala suggests that these neuronal systems may relate to the anxiety- or fear-regulating functions of the amygdala. Agonists at these receptors sites, when injected into the amygdala, reduce learned fear responses (Ben-Ari 1981; Fallon 1981).

The thalamus is another area that is important in fear responses. There is evidence from classically conditioned emotional responses that the processing of threatening stimuli involves the relay of sensory signals to the limbic forebrain directly from the thalamus and cortex (LeDoux et al. 1984, 1986a, 1986b). Fear responses to acoustic stimuli are disrupted by thalamic lesions, because responses depend on the connection between the thalamus and the amygdala. It has been speculated that the thalamic-amygdala pathways transmit rapidly a primitive representation of peripheral stimuli and could serve as preparation for the subsequent reception of processed inputs from the cortex (LeDoux 1987). The neurotransmitters involved in sensory transmission to the cortex in the limbic forebrain from the thalamus have not been elucidated. However, a recent study of the effects of electrical stimulation of the median and dorsal raphe nuclei on local cerebral glucose use in the rat suggests a relationship between serotonergic activity and thalamic function. Stimulation of the dorsal raphe nuclei produced an increase in glucose use in thalamic nuclei that subserve a processing of somatosensory, visual, and limbic information (Cudennec et al. 1988).

The hippocampus has been hypothesized to play a key position in anxiety development. The septohippocampal system has connections with limbic structures and cortical sensory areas. Lesions increase punished responses similar to the effects of anxiolytic drugs (Gray 1982). However, lesions of this area have had equivocal effects on classically conditioned fear and do not consistently affect the anxiolytic properties of antianxiety drugs (Gray 1982; LeDoux 1987). As mentioned above, high densities of 5-HT1A and 5-HT3 receptors are located in the hippocampus. It is tempting to speculate that the type of anxiety (e.g., generalized anxiety disorder) treated effectively by 5-HT$_{1A}$ agonists such as buspirone is mediated, in part, in the hippocampus.

There is also strong support, which has been reviewed in detail elsewhere, for the role of the locus coeruleus–norepinephrine system in the development of anxiety (Charney et al. 1984, 1987b). The locus coeruleus sends projections to other areas involved in fear or anxiety such as the thalamus, cerebral cortex, amygdala, hippocampus, and hypothalamus. In addition, the locus coeruleus receives afferents that suggest a critical role for this area in response to both the external and the internal environment (Moore and Bloom 1979). For example, alterations in blood pressure, in body temperature, and in distension of internal organs such as the bladder, colon, and stomach, all activate the locus coeruleus (Svennsson 1987). Other evidence supporting the role for the locus coeruleus in anxiety or fear includes the observation that drugs that activate the locus coeruleus are anxiogenic and drugs that decrease its function are anxiolytic (Charney et al. 1987b; Redmond 1979). In monkeys, stimulation of the locus coeruleus produces fearlike behavior, and lesions result in behavior consistent with reduced fear responses (Redmond 1979). In freely moving cats, threatening stimuli result in an activation of the locus coeruleus (Rasmussen and Jacobs 1986; Rasmussen et al. 1986). The activity of the locus coeruleus is highly regulated, with benzodiazepine, 5-HT, and opiate receptors having inhibitory effects, and vasoactive intestinal peptide (VIP), corticotropin-releasing factor (CRF), substance P, and acetylcholine resulting in activation (Charney et al. 1984, 1987b). The ability of 5-HT to decrease locus coeruleus firing suggests that interactions between noradrenergic and serotonergic systems may be relevant to the expression and treatment of anxiety or fear.

It is probable that the sensory processing areas of the neocortex linked with the limbic forebrain are particularly important in the interpretation of sensory stimuli from emotionally significant events such as those involving threat (LeDoux 1987). Electrical stimulation of serotonergic raphe nuclei produces well circumscribed increases in cortical glucose use in frontal motor, frontal sensorimotor, and frontoparietal somatosensory cortices. It has been suggested that this effect relates to the serotonergic innervation found during the ontogeny of the somatosensory cortex (Cudennec et al. 1988). The findings of high concentrations of 5-HT$_2$ receptors and the results of electrophysiological investigations of 5-HT$_2$ receptors in the cortex (see below) suggest that this receptor may be involved in the role of the neocortex in the translation of emotionally significant events.

## Laboratory Animal Behavioral Models of Anxiety and Fear: Relations to Serotonin Function

*Conflict Paradigms.* The standard conflict paradigm consists of training laboratory animals in operant tasks with food reward and then introducing short signal periods during which responding for food is rewarded but also punished by a mild electric shock. Animals stop responding during these short periods, exhibiting a so-called punishment-induced suppression. The punishment-induced blockade of ongoing behavior has been shown to be a useful method of identifying anxiolytic agents. Indeed, antianxiety drugs are the only group of agents consistently able to increase punished responding. In addition, there is a good correlation between the minimally effective antipunishment drug dose in animals and the average daily doses used in the treatment of anxiety. A variety of experimental paradigms have resulted in findings consistent with the hypothesis that depressed 5-HT neurotransmission is associated with attenuation of punishment-induced inhibition (Table 6-2) (Rasmussen and Aghajanian 1986; Soubrie 1986; Thiebot et al. 1982, 1983; Wise et al. 1972). In contrast, the administration of 5-HT agonists or electrical stimulation of the raphe nuclei does not consistently alter punished responding (Soubrie 1986).

*Novelty-Induced Inhibition.* Novel stimuli may increase the general level of fear. Often, but not invariably, one consequence of novelty is an inhibition of behavior, including inhibition of food intake in new surroundings, reduction of ambulation or exploratory behavior, and suppression of social interactions between a pair of rats. Antianxiety drugs generally release these forms of behavioral inhibition, thus

**Table 6-2.** Animal models of anxiety: relation to serotonergic function

| Model | Involvement of serotonergic neurons |
|---|---|
| Conflict paradigms | + + |
| Novelty-induced behavioral inhibition: | |
|   Hypophagia | + |
|   Exploratory Behavior | − |
| Nonreward-induced behavioral inhibition | + − |
| Fear-potentiated acoustic startle | − |

*Note.* + +, strong; +, moderate; + −, inconsistent; −, minimal.

increasing eating or drinking, exploratory behavior, and social inter-actions in unfamiliar situations (Gray 1982). 5-HT3 receptor an-tagonist drugs have been shown to have antianxiety potential, based upon social interaction tests (Jones et al. 1988). Destruction of serotonergic neurons, like the administration of benzodiazepines, preferentially increases social contacts in animals placed in unfamiliar situations. Serotonergic innervation of the lateral septum appears to be critically involved in this effect (Clarke and File 1982). Very few reports are available to suggest that depression of 5-HT transmission increases locomotor exploratory behavior in a novel situation (Beninger 1984; Clarke and File 1982; Soubrie 1986).

***Nonrewarding-Induced Inhibition.*** Analyzing animal behavior under nonreward conditions may shed further light on whether serotonergic neurons play an important role in the control of anxiety. This assumption derives from the concept that in animals, behavioral suppression under nonreward conditions is one possible adaptive response for avoiding frustration. Nonreward inhibition is highly sensitive to benzodiazepines. These drugs have been shown to in-crease the number of nonreward responses in the early trials of a nonrewarded session when the level of reward expectation is high. The effects of decreasing 5-HT transmission have been inconsistent in regard to bolstering nonreward-induced inhibition. From these data it has been suggested that one can extract the idea that a crucial dimension for the involvement of serotonergic neurons under condi-tions of nonreward is the animal's level of control over reward delivery or its level of reward expectancy (Soubrie 1986).

***Fear-Potentiated Startle Reflex.*** The fear-potentiated acoustic startle reflex has shown the potential to be a useful model of condi-tioned fear and the basis of a method for identifying anxiolytic drugs. The acoustic startle response, characterized by a distinctive, rapidly occurring pattern of whole-body muscular responses, is elicited by a sudden auditory stimulus. The magnitude of the startle response in rats can be increased by eliciting startle in the presence of a cue (e.g., a light) that was previously paired with a foot shock. Thus, after an initial training session, presentation of the light is a conditioned stimulus that presumably evokes a central fear state, which in turn produces an elevation in the acoustically elicited startle response (i.e., the fear-potentiated startle reflex). Consistent with a fear interpreta-tion, anxiolytic compounds such as amobarbital sodium (Chi 1965), diazepam (Berg and Davis 1984; Davis 1979a), or morphine (Davis 1979b) decrease the magnitude of fear-potentiated startle. In addition, fear-potentiated startle is also blocked by lesions of the central nucleus of the amygdala, a structure that has been implicated in fear in a variety of test situations (Hitchcock and Davis 1986).

Buspirone and gepirone, but not their common metabolite, 1-pyrimidinylpiperazine (1-PP), produced a dose-dependent reduction of fear-potentiated startle (Kehne et al. 1988). In evaluating the role of various neurotransmitter systems in mediating buspirone's blockade of fear-potentiated startle effect, lesions of the dorsal and median raphe nuclei or intraperitoneal injections of the 5-HT antagonists cinanserin or cyproheptadine did not alter fear-potentiated startle, nor did these treatments prevent buspirone from blocking fear-potentiated startle. The 5-HT$_{1A}$ agonist 8-hydroxy-2-(di-$n$-propylamino)tetralin (8-OH-DPAT) did not block fear-potentiated startle, even at doses that produced a marked "5-HT syndrome." Another 5-HT$_{1A}$ agonist, ipsapirone, blocked fear-potentiated startle only at a very high dose. $p$-Chlorophenylalanine and $p$-chloroamphetamine did not alter fear-potentiated startle. Taken together, the data do not support the hypothesis that the anxiolytic effects of buspirone are mediated by actions at 5-HT$_{1A}$ receptors, and these data more generally indicate that serotonergic neurons do not play an important role in fear-potentiated startle (Davis et al. 1988).

The reason for the apparent lack of involvement of 5-HT in potentiated startle relative to the other paradigms discussed above is currently not known. One plausible explanation has been provided by Soubrie (1986), who suggested that 5-HT may not be involved specifically in anxiety, but may more generally be involved in the inhibition of inappropriate responses. Thus, in animal models in which anxiety results in the animal's withholding responses to avoid punishment (i.e., not licking a water spout; not jumping off a platform onto an electrified grid), decreases in serotonergic function might generally diminish the ability of the animal to withhold that response. In the startle paradigm, the potentiation of an elicited reflex response, rather than the withholding of a response, serves as the index of fear (see Table 6-2).

### Effect of Antipanic and Antigeneralized Anxiety Drugs on Indices of Serotonin Function

*Nonbenzodiazepine Antipanic Drugs.* Neurophysiological studies have consistently indicated that a spectrum of drugs with antipanic (and antidepressant) properties (see later section) increase 5-HT neurotransmission with chronic administration (Blier et al. 1987; Charney et al. 1981; Heninger and Charney 1987). This increase appears to occur via at least two separate mechanisms. Specific 5-HT reuptake inhibitors, such as fluoxetine or fluvoxamine, and monoamine oxidase inhibitors (MAOIs) produce a decreased sensitivity of the 5-HT autoreceptor, whereas a spectrum of other drug treatments that include norepinephrine reuptake inhibitors such as desipramine and imipramine produce postsynaptic serotonergic supersensitivity by an as yet undetermined mechanism. This increase in 5-HT neurotransmission has not been associated with an increase in the density of 5-HT receptors. In contrast, 5-HT$_2$ receptors are reduced by long-term administration of most antipanic and antidepressant treatments.

The neurophysiological and binding studies of antipanic and antidepressant drug action must be considered in the context of the 5-HT receptor subtypes that have been evaluated and of the functional interaction between 5-HT$_1$ and 5-HT$_2$ receptors. As previously discussed, radioligand binding studies have provided evidence for the existence of distinct 5-HT$_1$ and 5-HT$_2$ binding sites. However, the range of functional properties of these receptors remains to be established. The 5-HT$_1$ receptor subtypes can be differentiated physiologically in the dorsal raphe nucleus, where 5-HT$_{1A}$, but not 5-HT$_{1B}$, ligands mimic the inhibitory action of 5-HT. It is the 5-HT$_{1A}$ receptor that is subsensitive after long-term treatment with 5-HT reuptake inhibitors and MAOIs. Radioligand binding studies that specifically evaluated changes in these somatodendritic 5-HT$_{1A}$ autoreceptors during antipanic and antidepressant treatment have found no alteration (Hensler and Frazer 1989).

In contrast to the dorsal raphe, where the 5-HT$_{1A}$ ligands are full agonists, in the hippocampus these compounds are partial agonists or antagonists. Neurophysiological investigations indicate that hippocampal 5-HT receptors become supersensitive after prolonged treatment with antipanic and antidepressant drugs. The enhancement of postsynaptic sites has been shown to occur in the lateral geniculate, amygdala, faciomotor nucleus, superchiasmatic nucleus, somatosensory cerebral cortex, and hippocampus. Radioligand binding studies have not been conducted that measure changes in these postsynaptic 5-HT$_{1A}$ receptors following antipanic treatment.

Neurophysiological studies indicate that in brain areas where 5-HT$_1$ and 5-HT$_2$ receptors exist, 5-HT$_2$ receptors reduce inhibitory effects and enhance excitatory effects produced by activation of the 5-HT$_1$ receptors. Of particular relevance is the finding in the prefrontal cortex that 5-HT$_2$ receptor activation works in opposition to the depressant effects of 5-HT on cell firing. Therefore, in the prefrontal cortex, the antipanic and antidepressant drug-induced decrease in 5-HT$_2$ receptors may be associated with a potentiation of the depressant effects of 5-HT on cell firing.

Behavioral paradigms of 5-HT function have been less helpful in assessing the effects of different drugs on 5-HT transmission in the brain. Most of the behavioral paradigms measure the function of 5-HT$_2$ receptors, which are distinct from the 5-HT$_1$ receptors studied neurophysiologically. In addition, behaviors assessed are primarily mediated in the brain stem and the spinal cord, and not the forebrain, which is most relevant to antipanic and antidepressant properties and where many of the neurophysiological studies have been conducted (Charney et al., in press).

*Antigeneralized Anxiety Drugs.* Benzodiazepine drugs and 5-HT$_{1A}$ receptor agonists, such as buspirone and gepirone, reduce 5-HT neurotransmission when given acutely. There is evidence that tolerance may develop to the 5-HT–reducing effects of benzodiazepines with chronic treatment (Collinge et al. 1983; Gonsalves and Gallager 1986; Lister and File 1983; Nutt and Cowen 1987).

One study has assessed the effect on 5-HT function of longer-term 5-HT$_{1A}$ receptor agonist treatment (Blier and de Montigny 1987). Gepirone acutely reduced the firing rate of dorsal raphe serotonergic neurons, which was followed by a gradual return to normal with 14 days of treatment. The responsivity of dorsal raphe neurons to lysergic acid diethylamide (LSD), 5-HT, 8-OH-DPAT, and gepirone, but not to GABA, was decreased after 14 days of gepirone treatment. A desensitization of the somatodendritic autoreceptor can thus account for the gradual recovery of the firing rate of the serotonergic neurons. This may provide an explanation for the delayed onset of action of drugs such as buspirone and gepirone in generalized anxiety disorder. From these data, and from the fact that postsynaptic 5-HT receptor function is unchanged by long-term gepirone administration, it has been hypothesized that 5-HT neurotransmission is increased by such a treatment. This increase may result from normal activity of serotonergic neurons and normal release of 5-HT combined with tonic activation by gepirone and presumably other 5-HT$_{1A}$ agonist drugs of normosensitive postsynaptic 5-HT receptors (Blier and de Montigny 1987) (Table 6-3).

## Summary of Preclinical Investigations

The neuroanatomical distribution of the serotonergic neuronal system is consistent with an important role in the regulation of anxiety or fear behavior. The presence of specific 5-HT receptors in brain regions such as the amygdala, thalamus, hippocampus, locus coeruleus, and neocortex suggests that specific sites exist for serotonergic regulation of anxiety. In particular, the high density of

5-HT$_{1A}$ receptors in the hippocampus indicates that this area may be important in the antianxiety actions of the 5-HT$_{1A}$ agonists such as buspirone and gepirone (Traber and Glaser 1987). The high concentration of 5-HT$_2$ receptors in the neocortex raises the possibility that effects in this region may account for the putative anxiolytic properties of 5-HT$_2$ receptor antagonists such as ritanserin (Ceulemans et al. 1985). 5-HT$_3$ receptors have been identified in the human amygdala and hippocampus, and behavioral studies suggest anxiolytic activity of 5-HT$_3$ receptor antagonists. The functional interactions between the serotonergic system and other neuronal systems, particularly the noradrenergic system, appear to be important, as evidenced by the regulation of locus coeruleus activity by 5-HT receptors.

The preclinical studies of behavioral models of anxiety and fear have provided an inconsistent picture of the relationship between 5-HT function and different forms of anxiety. The inconsistency is due, in part, to the differences that emerge between the effects of anxiolytic drugs and 5-HT depletion in the different models, and to the fact that it is neither sufficient nor necessary to involve serotonergic neurons in the production of anxiogenic effects. The probability that the different animal models are reflective of different types of anxiety or fear and that serotonergic neurons are involved in some, but not all, forms of anxiety contributes to the confusion.

The neurobiological investigations of the mechanisms of action of antipanic and nonbenzodiazepine antigeneralized anxiety drugs suggest that drugs whose anxiolytic effects relate to actions on 5-HT function (i.e., tricyclics, MAOIs, specific 5-HT reuptake inhibitors, 5-HT$_{1A}$ agonists) may be increasing net 5-HT function, a hypothesis

**Table 6-3.**  The net effect of antigeneralized anxiety and antipanic drugs on serotonin function

|  | Acute[a] | Chronic |
|---|---|---|
| Benzodiazepines | ↓ | 0 |
| 5-HT$_{1A}$ agonists (buspirone) | ↓ | ↑ ? |
| Tricyclics | 0 | ↑ |
| MAOIs | 0 | ↑ |
| Specific 5-HT reuptake inhibitors | ↑ | ↑ |

*Note.* ↓ = Decreased; 0 = no major effect; ↑ ? = increased in one study, but replication required. MAOIs = monoamine oxidase inhibitors.
[a]Acute (a few days or less) and chronic (at least a few weeks) refer to duration of drug administration.
*Source.* Data adapted from Blier and de Montigny (1987).

tantamount to a reversal of the original hypothesis that the ability of benzodiazepine drugs to reduce 5-HT activity was therapeutically important. The therapeutic mechanism of action of benzodiazepine drugs that are effective in the treatment of both panic disorder and generalized anxiety disorder, is probably due to actions on benzodiazepine-GABA receptors. Most of the current evidence suggests that benzodiazepine-induced changes in 5-HT function do not constitute the primary anxiolytic mechanism of action of benzodiazepine drugs.

## CLINICAL INVESTIGATIONS

### Studies of Serotonin Function in Patients With Anxiety Disorders

There have been relatively few clinical investigations of 5-HT function in patients with anxiety disorders. In one study designed to evaluate 5-HT function, the effects of intravenous tryptophan on prolactin secretion were determined in panic disorder patients and compared with these same effects in healthy subjects (Charney and Heninger 1986). The ability of tryptophan to increase prolactin levels, which has been shown to reflect 5-HT function, was not different between the patients and the healthy subjects.

Another method used to assess the possible role of 5-HT function in the development of panic anxiety has been the measurement of behavioral and biochemical responses to the 5-HT receptor agonist *m*-chlorophenylpiperazine (m-CPP). In a recently reported study (Charney et al. 1987a), m-CPP had anxiogenic effects in both the healthy subjects and the panic disorder patients. Panic attacks meeting DSM-III criteria occurred following m-CPP administration in approximately 45% of the patients and 30% of the healthy subjects (difference between the two groups is statistically nonsignificant) (Charney et al. 1987a). Other ratings of anxiety did not distinguish the two groups. m-CPP administration resulted in significant but similar increases in levels of cortisol, prolactin, and growth hormone in the healthy subjects and the patients. The results of this investigation suggested that serotonergic neuron dysfunction may not be of etiologic significance in most panic disorder patients. However, the observed anxiogenic properties of m-CPP suggest that additional studies of the role of systems in the pathophysiology of human anxiety disorders are indicated.

Low orally administered doses of m-CPP (0.25 mg/kg) have been reported to increase anxiety and produce panic attacks in panic disorder patients with or without major depression. In healthy control subjects and in patients with major depression but without panic disorder, the drug did not influence anxiety levels. The release of cortisol after m-CPP administration was also augmented in panic disorder patients. These findings have led to a hypothesis that hypersensitivity of postsynaptic 5-HT receptors may exist in some panic disorder patients (Kahn et al. 1988).

The normal prolactin responses to 5-HT agonists or precursors in patients with panic disorder are consistent with the findings of four of five studies, namely, that platelet imipramine binding is also normal in this diagnostic group (Innis et al. 1987; Lewis et al. 1985; Nutt and Fraser 1987; L. S. Schneider et al. 1987; Uhde et al. 1987). Radioactive imipramine has been shown to label a saturable high-affinity binding site in brain and platelet. Several lines of evidence suggest that this binding site is associated with, although not identical to, the 5-HT reuptake site.

One study (Norman et al. 1986) has measured platelet 5-HT uptake in patients with panic attacks, and found higher $V_{max}$ values in the patient group than in the control group, while the affinity constant ($K_m$) was not significantly different between groups. The authors suggested that there may be a specific abnormality of platelet 5-HT uptake in patients with panic attacks; furthermore, the increased platelet reuptake of 5-HT was consistent with a report of lower concentrations of this neurotransmitter in the plasma of panic disorder patients (P. Schneider et al. 1987). It should be noted, however, that the great bulk of plasma 5-HT is in platelets, and this 5-HT is believed to originate mainly from chromaffin cells in the gut wall. In another study, platelet 5-HT levels were found to be normal in patients with panic disorder compared to healthy subjects (Balon et al. 1987) (Table 6-4).

## Efficacy and Mechanism of Serotonin-Specific Drugs in Anxiety Disorders

There is emerging evidence that 5-HT reuptake-inhibiting drugs may have efficacy in the treatment of panic disorder. Both zimelidine and clomipramine have been shown to be effective antipanic agents in placebo-controlled investigations (Evans et al. 1986; Westenberg et al. 1987). In addition, fluvoxamine was shown to be effective in clinical studies comparing this agent to clomipramine and maprotiline (den Boer and Westenberg 1988). Fluoxetine has been reported to have efficacy in some panic disorder patients in an open study (Gorman et al. 1987).

The 5-HT$_{1A}$ agonists buspirone and gepirone have been shown to be effective antianxiety agents after several weeks of administration (Csanalosi et al. 1987; Goa and Ward 1986). The role for 5-HT in anxiolytic efficacy is also supported by work indicating that ritanserin, a 5-HT$_2$ receptor antagonist, has antianxiety properties. However, recent work indicates that neither buspirone nor ritanserin is an effective antipanic agent. This finding suggests that 5-HT$_{1A}$ and 5-HT$_2$ receptors may not be primarily involved in the pathogenesis of panic disorder (den Boer 1987; Sheehan et al. 1988). There are

**Table 6-4.** Clinical investigations of serotonergic function in panic disorder patients

| Paradigm | Dose | Method of administration | Finding |
|---|---|---|---|
| Tryptophan-induced prolactin response | 100 mg/kg | Intravenous | Prolactin response similar in patients and in healthy subjects |
| m-CPP-induced changes in behavior and neuroendocrine function | 0.1 mg/kg | Intravenous | Similar increases in levels of cortisol, prolactin, growth hormone, and anxiety in patients and in healthy subjects |
| | 0.25 mg/kg | Oral | Greater increases in levels of cortisol and anxiety in patients |
| Platelet imipramine binding | — | — | Similar in patients and in healthy subjects in four of five studies |
| Platelet 5-HT uptake and peripheral 5-HT levels | — | — | One study found increased $V_{max}$ in patients; one study reported decreased 5-HT in patients; another, no difference in platelet 5-HT |

*Note.* m-CPP = *m*-chlorophenylpiperazine.

ongoing studies evaluating the efficacy of 5-HT$_3$ receptor antagonists in generalized anxiety disorder. (Early oral presentations of the findings suggest that therapeutic properties may be present.)

There are few clinical studies assessing the effects of antianxiety drugs on 5-HT function. Both clomipramine and fluvoxamine have been demonstrated to enhance 5-HT function based upon their ability to enhance the prolactin response to intravenously administered tryptophan (Anderson and Cowen 1986; Price et al. 1989). In contrast, long-term alprazolam or diazepam treatment has no effect on this response, suggesting that there is no net increase or decrease in 5-HT function (Ceulemans et al. 1985; P. Schneider et al. 1987).

In healthy subjects, a single dose of diazepam attenuated the prolactin rise produced by tryptophan, suggesting that a reduction in 5-HT function occurs following acute, but not chronic, benzodiazepine treatment (Nutt and Cowen 1987). It has been speculated that benzodiazepine actions that become tolerant, such as sedation, are related to actions on serotonergic neurons rather than to antianxiety effects. There have been no clinical studies of the effects of 5-HT$_{1A}$ agonists on 5-HT function.

It is possible that some of the therapeutic effects of the 5-HT reuptake inhibitors are due to actions on other transmitter systems besides the serotonergic system. These compounds may decrease noradrenergic function, because 5-HT is inhibitory to the firing of noradrenergic neurons such as those of the locus coeruleus. For example, fluvoxamine has been shown to lower cerebrospinal fluid (CSF) levels of 3-methoxy-4-hydroxyphenylglycol (MHPG), a norepinephrine metabolite. In addition, some of the 5-HT inhibitors have been shown to downregulate β–adrenergic postsynaptic receptors in laboratory animals.

## CONCLUSIONS

This chapter has reviewed evidence suggesting possible roles for 5-HT neurotransmission in the development and treatment of anxiety states. While the preclinical neuroanatomical, neurophysiological, and behavioral studies suggest possible roles for 5-HT in anxiety or fear, the specific nature of the involvement has not been defined. This may be related to several factors, including the existence of 5-HT receptor subtypes with different functional properties, the number of behavioral models of anxiety that reflect different types of anxiety or fear, and the paucity of studies comparing the acute and chronic effects of anxiolytic drugs. The development and clinical availability of drugs with specific actions on 5-HT$_{1A}$, 5-HT$_2$, and 5-HT$_3$ receptors offer an opportunity to relate these receptors to different forms of human anxiety.

At present, clinical studies of the pathophysiology of anxiety disorders permit only preliminary observations. The ability of m-CPP to elicit anxiety is supportive of the hypothesis that serotonergic dysfunction may be anxiogenic. However, there is no consistent evidence of an abnormality in 5-HT function in either panic disorder or generalized anxiety disorder. The demonstration that 5-HT reuptake-inhibiting drugs have antipanic effects is not consistent with the hypothesis of hypersensitive 5-HT receptor function in panic disorder, because chronic administration of these drugs (e.g., when anxiolytic action occurs) appears to enhance 5-HT neurotransmission. The preliminary observation that the 5-HT$_2$ antagonist ritanserin does not have antipanic effects also fails to support this hypothesis. The function of the spectrum of 5-HT receptors, as well as the functional interaction between receptor subtypes, needs to be assessed in anxiety disorder patients.

The net effect of 5-HT$_{1A}$ agonists such as buspirone and gepirone on 5-HT function should be studied in humans. It needs to be determined whether the ability of these drugs to acutely reduce 5-HT firing is translated into reduced or increased 5-HT neurotransmission with chronic treatment when therapeutic actions are commencing. Since these drugs are effective for generalized anxiety disorder, but not panic disorder, patients, the possibility is raised that 5-HT dysfunction, particularly involving 5-HT$_{1A}$ receptors, may be more common in generalized anxiety disorder.

## *REFERENCES*

Anderson IM, Cowen PJ: Clomipramine enhances prolactin and growth hormone responses to L-tryptophan. Psychopharmacology (Berlin) 89:131–133, 1986

Balon R, Pohl R, Yeragani V, et al: Platelet serotonin levels in panic disorder. Acta Psychiatr Scand 75:315–317, 1987

Barnes JM, Barnes NM, Costall B, et al: Identification and characterization of 5-hydroxytryptamine$_3$ recognition sites in human brain tissue. J Neurochem 53:1787–1793, 1989

Ben-Ari Y: Transmitters and modulators in the amygdaloid complex: a review, in The Amygdaloid Complex. Edited by Ben-Ari Y. New York, Elsevier, 1981, pp 40–50

Beninger RJ: Effects of metergoline and quipazine on locomotor activity of rats in novel and familiar environments. Pharmacol Biochem Behav 20:701–705, 1984

Berg WK, Davis M: Diazepam blocks fear-enhanced startle elicited electrically from the brainstem. Physiol Behav 32:333–336, 1984

Blanchard DC, Blanchard RJ: Innate and conditioned reactions to threat in rats with amygdaloid lesions. J Comp Physiol Psychol 81:281–290, 1972

Blier P, de Montigny C: Modification of 5-HT neuron properties by sustained administration of the 5-HT$_{1A}$ agonist gepirone: electrophysiological studies in the rat brain. Synapse 1:470–480, 1987

Blier P, de Montigny C, Chaput Y: Modifications of the serotonin system by antidepressant treatments: implications for the therapeutic response in major depression. J Clin Psychopharmacol 7:24S–35S, 1987

Ceulemans DLS, Hoppenbrouwers M-LJA, Gelders YG, et al: The influence of ritanserin, a serotonin antagonist, in anxiety disorders: a double-blind placebo-controlled study versus lorazepam. Pharmacopsychiatry 18:303–305, 1985

Charney DS, Heninger GR: Serotonin function in panic disorders: the effect of intravenous tryptophan in healthy subjects and patients with panic disorder before and during alprazolam treatment. Arch Gen Psychiatry 43:1059–1065, 1986

Charney DS, Menkes DB, Heninger GR: Receptor sensitivity and the mechanism of action of antidepressant treatment: implications for the etiology and therapy of depression. Arch Gen Psychiatry 38:1160–1180, 1981

Charney DS, Heninger GR, Breier A: Noradrenergic function in panic anxiety: effects of yohimbine in healthy subjects and patients with agoraphobia and panic disorder. Arch Gen Psychiatry 41:751–763, 1984

Charney DS, Woods SW, Goodman WK, et al: Neurobiological mechanisms of panic anxiety: biochemical and behavioral correlates of yohimbine-induced panic attacks. Am J Psychiatry 144:1030–1036, 1987a

Charney DS, Woods SW, Goodman WK, et al: Serotonin function in anxiety. II. Effects of the serotonin agonist MCPP in panic disorder patients and healthy subjects. Psychopharmacology (Berlin) 92:14–24, 1987b

Charney DS, Price LH, Heninger GR: The receptor sensitivity hypothesis of antidepressant action: a synthesis and recommendations for future investigations. Arch Gen Psychiatry (in press)

Chi CI: The effect of amobarbital sodium on conditioned fear as measured by the potentiated startle response in rats. Psychopharmacologia 7:115–122, 1965

Clarke A, File SE: Selective neurotoxin lesions of the lateral septum: changes in social and aggressive behaviors. Pharmacol Biochem Behav 17:623–628, 1982

Cohen DH: The functional neuroanatomy of a conditioned response, in Neural Mechanisms of Goal-Directed Behavior and Learning. Edited by Thompson RF, Hicks LH, Shvyrkov B. New York, Academic, 1980

Collinge J, Pycock CJ, Taberner PV: Studies on the interaction between cerebral 5-hydroxytryptamine and gamma-aminobutyric acid in the mode of action of diazepam in the rat. Br J Pharmacol 79:637–643, 1983

Csanalosi I, Schweizer E, Case WG, et al: Gepirone in anxiety: a pilot study. J Clin Psychopharmacol 7:31–33, 1987

Cudennec A, Duverger D, Serrano A, et al: Influence of ascending serotonergic pathways on glucose use in the conscious rat brain. II. Effects of electrical stimulation of the rostral raphe nuclei. Brain Res 444:227–246, 1988

Davis M: Diazepam and flurazepam: effects on conditioned fear as measured with the potentiated startle paradigm. Psychopharmacology (Berlin) 62:1–7, 1979a

Davis M: Morphine and naloxone: effects on conditioned fear as measured with the potentiated startle paradigm. Eur J Pharmacol 54:341–347, 1979b

Davis M, Cassella JV, Kehne JH: Serotonin does not mediate anxiolytic effects of buspirone in the fear-potentiated startle paradigm: comparison with 8-OH-DPAT and ipsapirone. Psychopharmacology (Berlin) 94:14–20, 1988

den Boer JA: Serotonergic Mechanisms in Anxiety Disorders. The Hague, CIP Gegevens Koninklijke Biblioteek, 1987

den Boer JA, Westenberg HGM: Effect of a serotonin and noradrenaline uptake inhibitor in panic disorder: a double-blind comparative study with fluvoxamine and maprotiline. Int Clin Psychopharmacol 3:59–74, 1988

Downer JDC: Changes in visual gnostic function and emotional behavior following unilateral temporal lobe camage in the "split-brain" monkey. Nature 191:50–51, 1961

Evans L, Kenardy J, Schneider P, et al: Effect of a selective serotonin uptake inhibitor in agoraphobia with panic attacks. Acta Psychiatr Scand 73:49–53, 1986

Fallon JH: Histochemical characterization of dopaminergic, noradrenergic and serotonergic projections to the amygdala, in The Amygdaloid Complex. Edited by Ben-Ari Y. New York, Elsevier, 1981, pp 93–105

Goa KL, Ward A: Buspirone: a preliminary review of its pharmacological properties and therapeutic efficacy as an anxiolytic. Drugs 32:114–129, 1986

Gonsalves SF, Gallager DW: Tolerance to antipentylenetetrazol effects following chronic diazepam. Eur J Pharmacol 121:181–184, 1986

Gorman JM, Liebowitz MR, Fyer AJ, et al: An open trial of fluoxetine in the treatment of panic attacks. J Clin Psychopharmacol 7:329–332, 1987

Gray JA: Precis of "The neuropsychology of anxiety: an enquiry into the functions of the septo-hippocampal system." Behav Brain Sci 5:469–534, 1982

Heninger GR, Charney DS: Mechanism of action of antidepressant treatments: implications for the etiology and treatment of depressive disorders, in Psychopharmacology: The Third Generation of Progress. Edited by Meltzer HY. New York, Raven Press, 1987, pp 535–544

Hensler J, Frazer A: Effect of chronic antidepressant treatments on serotonin$_{1A}$ (5-HT$_{1A}$) receptor density and responsiveness. Neuroscience Abstracts 15:675, 1989

Hitchcock JM, Davis M: Lesions of the amygdala, but not of the cerebellum or red nucleus block conditioned fear as measured with the potentiated startle paradigm. Behav Neurosci 100:11–22, 1986

Innis RB, Charney DS, Heninger GR: Differential $^3$H-imipramine platelet binding in patients with panic disorder and depression. Psychiatry Res 21:33–41, 1987

Iwata J, LeDoux JE, Meeley MP, et al: Intrinsic neurons in the amygdaloid field projected to by the medial geniculate body mediate emotional responses conditioned to acoustic stimuli. Brain Res 383:195–214, 1986

Jones BJ, Costall B, Domeney AM, et al: The potential anxiolytic activity of GR 380 32F, a 5-HT$_3$ antagonist. Br J Pharmacol 93:985–993, 1988

Jones EG, Powell TPS: An experimental study of converging sensory pathways within the cerebral cortex of the monkey. Brain 93:793–820, 1970

Kahn R, Asnis G, Wetzler S, et al: Neuroendocrine evidence for a serotonin receptor supersensitivity in panic disorder. Psychopharmacology (Berlin) 96:360–364, 1988

Kapp BS, Frysinger RC, Gallagher M, et al: Amygdala central nucleus lesions: effects on heart rate conditioning in the rabbit. Physiol Behav 23:1109–1117, 1979

Kapp BS, Pascoe JP, Bixler MA: The amygdala: a neuroanatomical systems approach to its contributions to aversive conditioning, in The Neuropsychology of Memory. Edited by Butters N, Squire LR. New York, Guilford, 1984, pp 86–93

Kehne JH, Cassella JV, Davis M: Anxiolytic effects of buspirone and gepirone in the fear-potentiated startle paradigm. Psychopharmacology (Berlin) 94:8–13, 1988

LeDoux JE: Emotion, in Handbook of Physiology—The Nervous System V. Edited by Mountcastle VB, Plum F, Geiger SR. Bethesda, MD, American Physiological Society, 1987, pp 1–96

LeDoux JE, Sakaguchi A, Reis DJ: Subcortical efferent projections of the medial geniculate nucleus mediate emotional responses conditioned by acoustic stimuli. J Neurosci 4:683–698, 1984

LeDoux JE, Iwata J, Pearl D, et al: Disruption of auditory but not visual learning by destruction of intrinsic neurons in the medial geniculate body of the rat. Brain Res 371:395–399, 1986a

LeDoux JE, Sakaguchi A, Iwata J, et al: Interruption of projections from the medial geniculate body to an archi-neostriatal field disrupts the classical conditioning of emotional responses to acoustic stimuli in the rat. Neuroscience 17:615–627, 1986b

Lewis DA, Noyes R, Coryell W, et al: Tritiated imipramine binding to platelets is decreased in patients with agoraphobia. Psychiatry Res 16:1–9, 1985

Lister RG, File SE: Changes in regional concentrations in the rat brain of 5-hydroxytryptamine and 5-hydroxyindoleacetic acid during the development of tolerance to the sedative action of chlordiazepoxide. J Pharm Pharmacol 35:601–603, 1983

Mesulam MM, Vanhoesen G, Pandya DN, et al: Limbic and sensory connections of the IPL in the rhesus monkey. Brain Res 136:293–414, 1977

Moore RY, Bloom FE: Central catecholamine neuron systems: anatomy and physiology of the norepinephrine and epinephrine systems. Annu Rev Neurosci 2:113–168, 1979

Norman TR, Judd FK, Gregory M, et al: Platelet serotonin uptake in panic disorder. J Affective Disord 11:69–72, 1986

Nutt DJ, Cowen PJ: Diazepam alters brain 5-HT function in man: implications for the acute and chronic effects of benzodiazepines. Psychol Med 17:601–607, 1987

Nutt DJ, Fraser S: Platelet binding studies in panic disorder. J Affective Disord 12:7–11, 1987

Parent A, Descarries L, Beaudet A: Organization of ascending serotonergic systems in the adult rat brain: a radio autographic study after intraventricular administration of (3H)5-hydroxytryptamine. Neuroscience 6:115–138, 1981

Pazos A, Probst A, Palacios JM: Serotonin receptors in the human brain. III. Autoradiographic mapping of serotonin-1 receptors. Neuroscience 21:97–122, 1987a

Pazos A, Probst A, Palacios JM: Serotonin receptors in the human brain. IV. Autoradiographic mapping of serotonin-2 receptors. Neuroscience 21:123–139, 1987b

Peroutka SJ: 5-Hydroxytryptophan receptor subtypes: molecular, biochemical and physiological characterizations. Trends in Neuroscience 11:496–500, 1988

Price LH, Charney DS, Delgado PL, et al: Effects of desipramine and fluvoxamine treatment on the prolactin response to L-tryptophan: a test of the serotonergic function enhancement hypothesis of antidepressant action. Arch Gen Psychiatry (in press)

Rasmussen K, Aghajanian GK: Effect of hallucinogens on spontaneous and sensory-evoked locus coeruleus unit activity in the rat: reversal by selective 5-HT$_2$ antagonists. Brain Res 371:395–400, 1986

Rasmussen K, Jacobs BL: Single unit activity of locus coeruleus neurons in the freely moving cat. II. Conditioning and pharmacologic studies. Brain Res 371:335–344, 1986

Rasmussen K, Morilak DA, Jacobs BL: Single unit activity of locus coeruleus neurons in the freely moving cat. I. During naturalistic behaviors and in response to simple and complex stimuli. Brain Res 371:324–334, 1986

Redmond DE Jr: New and old evidence for the involvement of a brain norepinephrine system in anxiety, in Phenomenology and Treatment of Anxiety. Edited by Fann WE, Karacan I, Porkorny AD, et al. New York, SP Medical & Scientific Books, 1979, pp 153–203

Schneider LS, Munjack D, Severson JA, et al: Platelet [$^3$H]imipramine binding in generalized anxiety disorder, panic disorder, and agoraphobia with panic attacks. Biol Psychiatry 22:59–66, 1987

Schneider P, Evans L, Ross-Lee L: Plasma biogenic amine levels in agoraphobia with panic attacks. Pharmacopsychiatry 20:102–104, 1987

Sheehan DV, Raj AB, Sheehan KH, et al: The relative efficacy of buspirone, imipramine, and placebo: a preliminary report. Pharmacol Biochem Behav 29:815–817, 1988

Soubrie P: Reconciling the role of central serotonin neurons in human and animal behavior. Behav Brain Sci 92:319–364, 1986

Steinbusch HWM: Distribution of serotonin immunoreactivity in the central nervous system of the rat: cells bodies and terminals. Neuroscience 6:557–618, 1981

Svennsson TH: Peripheral, autonomic regulation of locus coeruleus noradrenergic neurons in brain: putative implications for psychiatry and pharmacology. Psychopharmacology (Berlin) 92:1–7, 1987

Swanson LW: The hippocampus and the concept of the limbic system, in Neurobiology of the Hippocampus. Edited by Seifert W. London, Academic, 1983, pp 23–40

Thiebot MH, Hamon M, Soubrie P: Attenuation of induced-anxiety in rats by chlordiazepoxide: role of raphe dorsalis benzodiazepine binding sites and serotonergic neurons. Neuroscience 7:2287–2294, 1982

Thiebot MH, Hamon M, Soubrie P: The involvement of nigral serotonin innervation in the control of punishment-induced behavioral inhibition in rats. Pharmacol Biochem Behav 19:225–229, 1983

Traber J, Glaser T: 5-HT$_{1A}$ receptor-related anxiolytics. Trends in Pharmacological Sciences 8:432–437, 1987

Uhde TW, Berrettini WH, Roy-Byrne PP, et al: Platelet [3H]imipramine binding in patients with panic disorder. Biol Psychiatry 22:52–58, 1987

Waeber C, Hoyer D, Palacios JM: 5-Hydroxytryptamine$_3$ receptors in the human brain: autoradiographic visualization using [3H]ICS 205-930. Neuroscience 31:393–400, 1989

Westenberg HGM, den Boer JA, Kahn RS: Psychopharmacology of anxiety disorders: on the role of serotonin in the treatment of anxiety states and phobic disorders. Psychopharmacol Bull 23:145–149, 1987

Wise CD, Berger BD, Stein L: Benzodiazepines: anxiety reducing activity by reduction of serotonin turnover in the brain. Science 177:180–183, 1972

# Chapter 7

## *Serotonin in Eating Disorders*

Timothy D. Brewerton, M.D.
Harry A. Brandt, M.D.
Michael D. Lessem, M.D.
Dennis L. Murphy, M.D.
David C. Jimerson, M.D.

# Chapter 7

## *Serotonin in Eating Disorders*

Anorexia nervosa is an eating disorder that occurs predominately in young, occidental white females. (See Table 7-1 for DSM-III-R diagnostic criteria for this disorder.) Its incidence appears to have been on the rise during the last few decades (Jones et al. 1980; Kendell et al. 1973; Rosenzweig and Spruill 1987; Szmukler et al. 1986; Theander 1970, 1985). Anorexia nervosa is associated with substantial morbidity and mortality (Ferguson 1985; Harris 1983; Jacobs and Schneider 1985; Mitchell et al. 1983) and can be one of the most treatment-resistant psychiatric illnesses (Agras and McCann 1987; Hsu et al. 1979; Morgan and Russell 1975; Swift et al. 1987). Generally, anorexic patients can be classified into two basic subtypes, "restricter" patients and "bulimic" patients, based on the absence or presence of bulimic behaviors (i.e., binge eating, vomiting, and/or laxative abuse). If bulimic behaviors are present, the DSM-II-R criteria for bulimia nervosa may also be met. Bulimic anorexic patients appear to be more prone to have depressive illness and to engage more frequently in stealing, drug abuse, and autoaggressive behaviors (Beumont et al. 1976b; Casper et al. 1980; Garfinkel et al. 1980; Strober 1981). As will be discussed later in this chapter, these basic subtypes of anorexia nervosa patients also exhibit psychobiological characteristics suggestive of differences in central serotonin (5-hydroxytryptamine [5-HT]) function.

While the term "bulimia" often refers simply to the presence of binge eating and its associated counteractive measures, in DSM-III (American Psychiatric Association 1980) it refers to a specific psychiatric syndrome that is exclusive of anorexia nervosa. Because bulimic symptoms clearly occur in the context of anorexia nervosa, this has understandably created confusion in the nomenclature as well as in the conceptualization of the phenomenon as a disorder. Bulimic symptoms also occur in the context of normal weight (Boskind-Lodahl 1976; Crisp 1981; Palmer 1979; Russell 1979) and obesity (Kornhaber 1970; Loro and Orleans 1981; Marcus and Wing 1987; Stunkard 1959), and can be either primary or secondary to a number of medical conditions or drug-induced states. Bulimia was originally described in the context of anorexia nervosa, first by Richard Morton and later by others (Casper 1983). In 1979 Russell coined the term "bulimia nervosa" to describe "an ominous variant of anorexia nervosa" occurring in both low- and normal-weight patients and characterized by intractable urges to eat, counteractive behaviors against resultant weight gain, and a morbid dread of fatness (Russell 1979). The American Psychiatric Association has now incorporated Russell's terminology into DSM-III-R criteria (Table 7-2). Patients with DSM-III-R bulimia nervosa need not have had a prior history of anorexia nervosa, as recommended by Russell (1985) in his most recently revised criteria for the disorder. For the purposes of this chapter, bulimia nervosa will refer to the disorder as defined by DSM-III-R criteria.

## SEROTONIN AND THE PHENOMENOLOGY OF THE EATING DISORDERS

### Serotonin and Feeding

Manipulations of central 5-HT function in animals and in man result in marked changes in feeding behaviors, particularly satiety responses

**Table 7-1.**  Anorexia nervosa: DSM-III-R diagnostic criteria

---

A. Refusal to maintain body weight over a minimal normal weight for age and height, e.g., weight loss leading to maintenance of body weight 15% below that expected; or failure to make expected weight gain during period of growth, leading to body weight 15% below that expected.

B. Intense fear of gaining weight or becoming fat, even though underweight.

C. Disturbance in the way in which one's body weight, size, or shape is experienced, e.g., the person claims to "feel fat" even when emaciated, believes that one area of the body is "too fat" even when obviously underweight.

D. In females, absence of at least three consecutive menstrual cycles when otherwise expected to occur (primary or secondary amenorrhea). (A woman is considered to have amenorrhea if her periods occur only following hormone, e.g., estrogen, administration.)

---

*Source.* Reproduced, with permission, from American Psychiatric Association *Diagnostic and Statistical Manual of Mental Disorders*, 3rd Edition, Revised. Washington, D.C., American Psychiatric Association, 1987, p. 67. Copyright 1987, American Psychiatric Association.

(Blundell 1977, 1984, 1986; Samanin et al. 1982; Silverstone and Goodall 1986). Pharmacologic enhancement of 5-HT neurotransmission generally leads to increased satiety, as evidenced by reductions in meal size, rate of eating, and body weight (Blundell 1984, 1986), while the latency to the onset of feeding remains unaffected (Rogers and Blundell 1979). 5-HT agonists, particularly fenfluramine, are reported to decrease specifically carbohydrate consumption while sparing protein intake (Leibowitz and Shor-Posner 1986; Li and Anderson 1984; Wurtman and Wurtman 1986; Wurtman et al. 1981). An important exception is the induction of feeding by 5-HT$_{1A}$ receptor activation (Dourish et al. 1986). Attenuation of 5-HT neurotransmission generally leads to decreased satiety and increased food consumption and weight gain (Blundell 1984, 1986; Garattini et al. 1988). The predominant effects of 5-HT on feeding are thought to be mediated centrally via the medial hypothalamus (Leibowitz and Shor-Posner 1986).

## Serotonin and Fasting

Short-term dieting can cause upregulation of platelet 5-HT receptors (Goodwin et al. 1987c), as well as increased neuroendocrine responsivity to intravenously administered L-tryptophan, particularly in women (Cowen and Charig 1987; Goodwin et al. 1987b). These data are relevant given that the severe restriction of food intake common

**Table 7-2.**   Bulimia nervosa: DSM-III-R diagnostic criteria

A.  Recurrent episodes of binge eating (rapid consumption of a large amount of food in a discrete period of time).

B.  A feeling of lack of control over eating behavior during the eating binges.

C.  The person regularly engages in either self-induced vomiting, use of laxatives or diuretics, strict dieting or fasting, or vigorous exercise in order to prevent weight gain.

D.  A minimum average of two binge eating episodes a week for at least three months.

E.  Persistent overconcern with body shape and weight.

*Source.* Reproduced, with permission, from American Psychiatric Association *Diagnostic and Statistical Manual of Mental Disorders*, 3rd Edition, Revised. Washington, D.C., American Psychiatric Association, 1987, pp. 68–69. Copyright 1987, American Psychiatric Association.

to bulimia nervosa patients typically precedes binge-eating behavior (Mitchell et al. 1986) and also probably contributes to mood lability. These changes in 5-HT receptor sensitivity could conceivably be involved in the inherently rewarding aspects of dieting that are experienced by eating-disorder patients.

## Serotonin and Mood

Further indirect evidence for abnormal serotonergic activity in patients with bulimic symptoms derives from several striking similarities between bulimia and depression. Bulimic patients have been noted by several authors not only to have a high frequency of depressed mood (Abraham and Beumont 1982; Brewerton et al. 1986a; Hatsukami et al. 1984; Johnson and Larson 1982; Weiss and Ebert 1983 but also to have lifetime histories of major depression of approximately 70% (Hudson et al. 1984, 1987; Walsh et al. 1985).

## Serotonin and Anxiety

Anxiety, a common symptom reported by patients with bulimia nervosa (George et al. 1987), is also modulated by 5-HT, although its role is a complex one (Charney et al. 1987; Davis et al. 1985; Gardner 1985; Johnston and File 1986). Most animal studies indicate that enhanced 5-HT function generally increases anxiety, a phenomenon probably mediated through activation of receptors other than those of 5-HT$_{1A}$ subtype (Carli and Samanin 1988; Engel et al. 1984; Glaser and Traber 1985; Peroutka 1985). In particular, enhancement of 5-HT function inhibits punishment-suppressed behavior, whereas attenuation leads to the release of punishment-suppressed behavior. Such a functional 5-HT deficit in bulimic patients could conceivably contribute to the loss of control characteristic of bulimic behaviors such as binge eating, purging, stealing, sexual acting out, and various forms of self-destructive or autoaggressive behavior. The high frequency of obsessional symptoms (Cooper and Fairburn 1986; Hudson et al. 1987; Rothenberg 1986) reported in bulimic patients may also involve serotonergic dysfunction (Zohar and Insel 1987).

## Serotonin, Impulsivity, and Aggression

Impulsivity, a phenomenon related to the release of punishment-suppressed behavior discussed above, is modulated by 5-HT neurotransmission (Linnoila et al. 1983). Some forms of aggression toward self and others can be conceptualized as disinhibited behavior (Åsberg et al. 1987). Muricidal activity in rats is induced by a diet low in L-tryptophan as well as by administration of $p$-chlorophenylalanine (p-CPA), an inhibitor of 5-HT synthesis, and this activity is reversed by orally administered L-tryptophan (Valzelli et al. 1981). Lower levels of cerebrospinal fluid (CSF) 5-hydroxyindoleacetic acid (5-HIAA) in humans have been associated with suicide attempts, especially by violent means (Åsberg et al. 1976a, 1976b; Brown et al. 1982). The behavior of bulimia nervosa patients is often of an impulsive nature—for example, binge eating, suicide attempts, alcohol and/or drug abuse, stealing, and increased sexual behavior (Crisp et al. 1980; Garfinkel and Garner 1982; Russell 1979). Interestingly, several authors have noted the association between disturbed eating behaviors and self-mutilation (French and Nelson 1972; Goldney and Simpson 1975; Rosenthal et al. 1972; Simpson 1975).

### Serotonin and Alcoholism

The pathophysiological role of 5-HT in the development and maintenance of alcoholism is discussed elsewhere (Ballenger et al. 1979; Myers and Melchior 1978; Naranjo et al. 1986; Roy et al. 1987). Bulimia nervosa can be conceptualized as an addiction to food, much like an addiction to alcohol (Scott 1984). Biochemically, alcohol intoxication increases CSF 5-HIAA levels, which correlate to blood ethanol concentrations (Borg et al. 1985); this increase is not unlike the increase in central 5-HT following food ingestion in bulimic patients (Kaye et al. 1988a). Recent studies show that 5-HT uptake inhibitors decrease alcohol intake (Murphy et al. 1985; Naranjo et al. 1984, 1986, 1987), food intake (Gill and Amit 1987), and bulimic symptoms (Freeman et al. 1988). Alcohol abuse is quite common in bulimia nervosa patients (Mitchell et al. 1985) and their families (Bulik 1987; Hudson et al. 1987). Both alcoholic patients and bulimic patients may be seeking to modify a serotonergic defect by superficially different, yet probably similar psychobiologic mechanisms.

### Serotonin and Other Neurochemicals

Serotonin is a major modulator of several hormones that have been reported to be dysfunctional in eating disorder patients. These hormones include corticotropin-releasing hormone (CRH), cortisol, luteinizing hormone (LH), follicle-stimulating hormone (FSH), and vasopressing. Evidence in animals and humans also attests to the interdependence of the serotonergic and the adrenergic systems in that an alteration in either one can produce an alteration in the other. A comprehensive discussion of these topics is beyond the scope of this chapter.

## EVIDENCE FOR SEROTONERGIC DYSREGULATION IN ANOREXIA NERVOSA

### Peripheral Studies

Coppen and colleagues (1976) were the first to report a reduction in plasma L-tryptophan levels, both bound and free forms, in anorexia nervosa patients. Upon refeeding, levels of the free, but not the bound, form of plasma L-tryptophan remained reduced. Johnston and associates (1984) also found reduced plasma L-tryptophan levels as well as a reduced L-tryptophan/neutral amino acid (NAA) ratio in anorexic patients, thereby implying that there was reduced availability of L-tryptophan for 5-HT synthesis. However, Russell (1967) reported that plasma L-tryptophan concentration was normal in anorexic patients. Likewise, Kaye and coworkers (1984b) found no differences in the L-tryptophan/NAA ratio between anorexic patients before or after short- or long-term weight recovery and control subjects. Although the hypothesis is theoretically sound, it is not established that a reduced L-tryptophan/NAA ratio reflects reduced central 5-HT function. Even though L-tryptophan administration is known to increase CSF 5-HIAA levels in humans (Dunner and Goodwin 1972; Eccleston et al. 1970; Koskiniemi et al. 1985), neither the plasma concentration of L-tryptophan nor the L-tryptophan/NAA ratio has been shown to be correlated with CSF 5-HIAA levels (Bjerkenstedt et al. 1985; Hagenfeldt et al. 1984).

Urinary excretion of 5-HIAA has been reported to be significantly decreased in anorexic patients in comparison to control subjects (Riederer et al. 1982). However, because refeeding resulted in a return to normal concentrations in the four recovered patients studied, it is probable that this decrease in urinary excretion of 5-HIAA was a result of reduced nutritional intake.

Platelet monoamine oxidase (MAO) activity has been studied in a variety of psychiatric disorders, although recent data indicate that platelet MAO activity is poorly correlated with brain MAO activity (Young et al. 1986). Nevertheless, Biederman and coworkers (1984b) reported that platelet MAO activity was decreased in anorexic patients with major depression in comparison to non-depressed anorexic patients and control subjects. The data were not separated into bulimic and nonbulimic groups for analysis.)

Platelet imipramine binding and 5-HT uptake have been used, particularly in affective illness, in an attempt to identify defects in 5-HT transport mechanisms (Healy and Leonard 1987). Available data indicate that nondepressed patients with anorexia nervosa have decreased imipramine binding (Weizman et al. 1986). However,

5-HT uptake has been reported to be no different in anorexic patients in comparison to control subjects in two studies (Weizman et al. 1986; Zemishlany et al. 1987). None of these studies reported on the bulimic status of the anorexic patients studied. Evidence regarding defects in 5-HT transport mechanisms is therefore too sparse for any firm conclusion to be made. Future studies are needed in bulimic and nonbulimic anorexic patients of low weight and after complete nutritional rehabilitation, given that the peripheral measures of 5-HT function noted above, as well as the basal concentrations of CSF 5-HIAA noted below, appear to normalize after refeeding.

## Cerebrospinal Fluid Studies

Concentrations of CSF 5-HIAA have been reported to be lower in low-weight female anorexic patients when compared to these same patients after weight restoration (Kaye et al. 1984b) (Figure 7-1) or to healthy control women (Gillberg 1983; Kaye et al. 1988b). However, Gerner and colleagues (1984) found no differences between anorexic patients and control subjects in either CSF 5-HIAA or CSF L-tryptophan concentrations. It is notable that patients in the latter study were somewhat heavier than those in the former study. In addition, the patients in the Gerner et al. study had already undergone some degree of refeeding and weight gain. These authors' finding is compatible with normalization of CSF 5-HIAA concentrations following refeeding (Kaye et al. 1984b, 1988b) (Figure 7-1). In summary, the reduction in basal CSF 5-HIAA levels in underweight patients with anorexia nervosa, regardless of the concurrent presence of bulimia nervosa, appears to be state-dependent and a result of starvation. In contrast, after probenecid administration, lower CSF 5-HIAA concentrations have been reported in weight-recovered bulimic anorexic patients compared to restricter anorexic patients (Kaye et al. 1984a). This finding may represent a trait-related decrease in 5-HT turnover, independent of the effects of starvation, that is specific to patients with bulimia nervosa. Alternatively, this lower concentration could represent a residual change from the low-weight episode that is more persistent in bulimic patients.

## Neuroendocrine Studies

The pharmacologic challenge strategy has become increasingly recognized as a safe and viable method of investigating the functional status of central 5-HT (Mueller et al. 1985, 1986; Murphy et al. 1986). At the National Institute of Mental Health our group has been studying prolactin responses to L-tryptophan (100 mg/kg iv) and *m*-chlorophenylpiperazine (m-CPP; 0.5 mg/kg po), a specific 5-HT

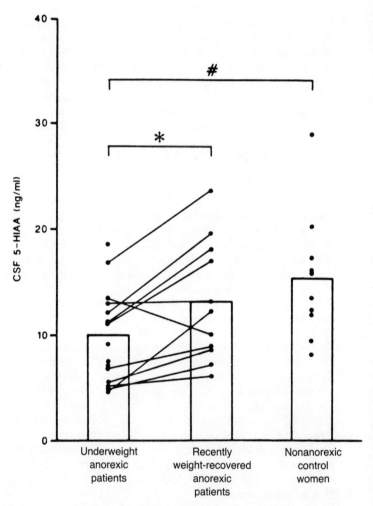

**Figure 7-1.** Individual data points and mean concentrations of 5-hydroxy-indoleacetic acid (5-HIAA) in CSF for anorexic patients and nonanorexic control subjects. Asterisk (*) indicates a significant ($P < .01$) difference between underweight and recently recovered anorexic patients; (#) indicates a significant ($P < .01$) difference between underweight anorexic patients and control subjects. Reproduced from Kaye et al., *Biological Psychiatry* 23:102–105, 1988b, with permission of Elsevier Science Publishing Company. Copyright 1988, Elsevier Science Publishing Company.

receptor agonist (Brewerton et al. 1986b, 1987a, 1987b, 1988, submitted for publication). Female patients (age = 25.6 ± 5.9 years) with DSM-III–defined anorexia nervosa were studied before (35.7 ± 4.6 kg; 61.7 ± 7.0% average body weight [ABW]) and 4 weeks after weight restoration to a predetermined goal weight (49.0 ± 2.8 kg; 85.7 ± 3.0% ABW). (All patients also retrospectively met DSM-III-R criteria for anorexia nervosa.) Healthy female control subjects (age = 26.4 ± 5.1 years) were selected on the basis of a negative lifetime history of any psychiatric disorder in themselves or in first-degree relatives as determined by a SADS-L interview (Endicott and Spitzer 1978). At the time of the study and in the past, the weights of all control subjects (57.4 ± 6.1 kg; 99.2 ± 9.9% ABW) and of their first-degree relatives had been between 85% and 120% of ABW as determined by standardized weight tables (Society of Actuaries 1980). Active drugs or placebo was given on separate days at least 48 hours apart using a randomized, double-blind design. All subjects were healthy (with the exception of malnutrition in the underweight patients) as determined by history, physical examination, ECG, and laboratory tests. In addition, all subjects were medication-free for at least 4 weeks prior to the beginning of the study.

The mean peak change in prolactin response (peak minus baseline) to L-tryptophan was significantly blunted in 12 low-weight patients (6.4 ± 4.8 ng/ml, $P < .05$, Mann-Whitney U test) and in 10 goal weight patients (11.0 ± 6.5 ng/ml, $P < .05$) compared to 16 control subjects (25.4 ± 21.0 ng/ml) (Figure 7-2). In 10 paired anorexic patients there was a trend for the patients to have higher peak delta prolactin responses at goal weight than at low weight ($P \leq .09$, paired Student's t test). Likewise, peak delta prolactin responses to m-CPP were blunted in 12 low-weight patients (3.3 ± 3.0 ng/ml, $P < .005$, Mann-Whitney U test) and in nine goal-weight patients (7.2 ± 6.5 ng/ml, $P < .02$, Mann-Whitney U test) versus responses in 15 control subjects (27.4 ± 28.7 ng/ml) (Figure 7-2). In the eight paired anorexic patients there was a trend, similar to that of the L-tryptophan findings, for goal-weight responses to be higher than the low-weight responses ($P < .1$, Mann-Whitney U test). These results could not be accounted for by differences in plasma drug concentrations, baseline estradiol levels, Hamilton Rating Scale or Beck Depression Inventory scores, or number of years ill. Peak delta prolactin responsive follow- ing m-CPP administration was significantly correlated with CSF 5-HIAA obtained from lumbar punctures 3 to 10 days prior to challenge with m-CPP in the low-weight anorexic patients ($n = 12$, r = −.69, $P \leq .009$), but not in the goal-weight anorexic patient (M. Lesem et al., unpublished data). No significant correlations were found between peak delta prolactin response following L-tryptophan and CSF 5-HIAA administration in either group of anorexic patients. The possible reasons for the presence of a relationship in the low-weight anorexic patients between CSF 5-HIAA and the prolactin response to m-CPP, but not to L-tryptophan, are not well understood, but they may relate to differential involvement of pre- and postsynaptic mechanisms, 5-HT receptor subtypes, and anatomic loci of action.

In summary, responsivity in hypothalamo-pituitary serotonergic pathways at or distal to the postsynaptic 5-HT receptor mediating prolactin responses is blunted in patients with anorexia nervosa whether at low weight or after attainment of goal weight. Although there appeared to be a move toward normalization of receptor sensitivity after nutritional rehabilitation, peak delta prolactin responses to both L-tryptophan and m-CPP remained blunted in the

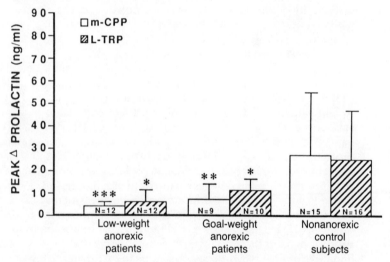

**Figure 7-2.** Mean peak change in prolactin responses ($\pm$ SD) to *m*-chlorophenylpiperazine (m-CPP) and L-tryptophan (L-TRP) in anorexic patients at low weight and at goal weight and in nonanorexic control subjects. (***) indicates $P < .005$, Mann-Whitney U test, low-weight anorexic patients vs. nonanorexic control subjects after m-CPP administration; (**) indicates $P < .02$, Mann-Whitney U test, goal-weight anorexic patients vs. control subjects after m-CPP; (*) indicates $P < .05$, Mann-Whitney U test, low-weight anorexic patients vs. control subjects and goal-weight anorexic patients after L-TRP.

refed patients in comparison to control subjects. It is unlikely that the blunting is due to deficient prolactin stores or to an impairment in the lactotroph itself, since peak prolactin responses to thyrotropin-releasing hormone (TRH) in low-weight anorexic patients have been reported to be normal in most (Beumont et al. 1976a; Kiriike et al. 1987; Macaron et al. 1978; Vigersky et al. 1976), but not all (Waldhauser et al. 1984), studies. No conclusions can be made about differences in receptor sensitivity between bulimic anorexic patients and restricter anorexic patients at this time given the small number of restricter patients in our sample (two at low weight, one at goal weight). Within the group of low-weight anorexic patients, increased central 5-HT turnover appears to be associated with blunted prolactin responses following m-CPP (but not L-tryptophan) administration. The reasons for the disappearance of this relationship after refeeding are unknown, but probably involve nutritional factors.

## Pharmacologic Studies

The pharmacologic treatment of anorexia nervosa, which has been reviewed elsewhere (Garfinkel and Garner 1987; Gwirtsman et al. 1984), has been largely disappointing. Although lithium, cyproheptadine, amitriptyline, and clomipramine have all shown some therapeutic efficacy in double-blind trials, it is notable that all these studies have been conducted in unison with structured treatment programs using nonpharmacologic methods. Nevertheless, it is pertinent to this discussion that these agents alter 5-HT function in a variety of ways. Eight patients receiving lithium gained significantly more weight during weeks 3 and 4 of treatment than eight patients receiving placebo (Gross et al. 1984). Notably, most of the anorexic patients in this study appeared to be of the bulimic type, which is in keeping with reports of lithium's positive effect in normal-weight patients with bulimia nervosa (Hsu 1984) (see below). Lithium enhances and stabilizes 5-HT neurotransmission in complex and incompletely understood ways involving both peripheral and central mechanisms (Blier et al. 1987; Brewerton and Reus 1983; Glue et al. 1986; Knapp and Mandell 1973).

Early studies that led to the use of the 5-HT receptor antagonist cyproheptadine in anorexia nervosa have been reviewed by Goldbloom (1987). In an important study by Halmi and coworkers (1986), cyproheptadine significantly increased treatment efficiency in nonbulimic or restricter anorexic patients and significantly impaired treatment efficiency in bulimic anorexic patients in comparison to amitritypline- or placebo-treated groups. Both cyproheptadine and amitriptyline were moderately better than placebo for the total group of anorexic patients in terms of antidepressant response and weight gain. This study provided pharmacologic evidence for a difference in 5-HT function between bulimic and nonbulimic anorexic patients (also suggested by differences in CSF 5-HIAA concentrations after probenecid administration [see below]), although the exact nature of this difference remains unknown. Cyproheptadine blocks 5-HT receptor activation, presumably at the level of the hypothalamus, to decrease satiety. However, cyproheptadine and other 5-HT antagonists appear to downregulate postsynaptic 5-HT receptors in animals (Blackshear et al. 1983, 1986; Gandolfi et al. 1985), an effect opposite from that expected. Regardless of this latter effect, cyproheptadine has been shown to decrease electrical activity in the "satiety center" (i.e., ventromedial hypothalamus) (Chakrabarty et al. 1967; Oomura et al. 1973) and increase electrical activity in the "feeding center" (i.e., lateral hypothalamus) of animals (Oomura et al. 1973). An earlier study by Vigersky and Loriaux (1977) found no significant difference in response between cyproheptadine-and placebo-treated groups, but the patients were not strictly divided into bulimic and nonbulimic groups for analysis. However, patients in the cyproheptadine group were older, had been ill longer, and had a much higher frequency of bulimic behaviors (vomiting and abuse of laxatives, diuretics, and/or drugs) than the patients in the placebo group, who had a higher frequency of nonbulimic behaviors (abstinence and/or exercise). A negative amitriptyline versus placebo study by Biederman and colleagues (1984a) also did not examine the bulimic and nonbulimic data separately.

Lacey and Crisp (1980) reported that weight-recovered anorexic patients receiving clomipramine had more stable eating habits and were better able to maintain their weights than were the group of patients receiving placebo. This is a major finding given that many anorexic patients tend to lose weight promptly upon discharge from inpatient facilities, largely due to unremitting obsessional concerns about obesity. The therapeutic efficacy of clomipramine in obsessive-compulsive disorder, which shares many clinical features with anorexia nervosa, has been related to its potent effects on 5-HT neurotransmission (Zohar and Insel 1987). Further pharmacologic studies on the prevention of relapse in weight-recovered anorexic patients using clomipramine and other serotonergic drugs are warranted.

## EVIDENCE FOR SEROTONERGIC DYSREGULATION IN BULIMIA NERVOSA

### Peripheral Studies

Although L-tryptophan/NAA ratios in normal-weight patients with bulimia nervosa are not significantly different from those in healthy control subjects (Lydiard et al. 1988), the L-tryptophan/NAA ratio has been shown to vary inversely with binge severity (Kaye et al. 1988a). The $V_{max}$ of platelet 5-HT uptake was reportedly higher in 26 normal-weight patients with bulimia nervosa in comparison to 21 age-, weight-, and sex-matched control subjects (Goldbloom et al. 1988). Because only three of these patients had concurrent major depressive illness, these data suggest an alteration in 5-HT uptake specific to bulimia nervosa. This is further substantiated by the lack of correlation between Hamilton depression ratings and $V_{max}$ of platelet 5-HT uptake. There are no published reports of platelet imipramine binding in patients with bulimia nervosa.

## Cerebrospinal Fluid Studies

There are no published reports of CSF 5-HIAA concentrations in bulimia nervosa. However, Kaye and coworkers (personal communication, 1989) and Lesem and coworkers (personal communication, 1989) have not found significant differences in baseline CSF 5-HIAA concentrations in two separate groups of normal-weight bulimic patients in comparison to control subjects. Nonetheless, it is conceivable that any differences in CSF 5-HIAA might not be manifest without probenecid pretreatment (see above). Future studies might also benefit from direct measurement of CSF 5-HT, which has been reported to be increased in depressed patients (Gjerris et al. 1987).

## Neuroendocrine Studies

Our group has studied normal-weight patients with DSM-III bulimia (age = $24.7 \pm 4.8$ years; $54.7 \pm 4.9$ kg; $57.5 \pm 6.3\%$ ABW) using the same challenge paradigm as that described for anorexia nervosa patients (see above) (Brewerton et al. 1987a, 1987b, submitted for publication). (All patients also retrospectively met DSM-III-R criteria for bulimia nervosa.) Except for six normal-weight outpatients, all patients had been hospitalized and had abstained from binge eating and vomiting for at least 3 weeks. All inpatients had also been on controlled diets such that their weights were stable for at least 3 weeks prior to the beginning of the study.

The peak delta prolactin response following m-CPP administration in 26 bulimic patients ($6.5 \pm 5.9$) ng/ml) was significantly lower than that of 15 control subjects ($27.3 \pm 28.7$ ng/ml, $P < .002$, Mann-Whitney U Test) (Figure 7-3). Although the 10 bulimic patients with concurrent major depression had lower peak delta prolactin levels (4.2

± 4.1 ng/ml) than the 16 bulimic patients without this diagnosis (8.0 ± 6.5 ng/ml), this difference did not reach statistical significance ($P$ = .11, unpaired Student's t test) (Figure 7-4). Bulimic patients without major depression had significantly lower peak delta prolactin levels than did control subjects ($P < .02$, Mann-Whitney U Test). These group differences could not be attributed to age, weight, %ABW, peak m-CPP concentrations, Hamilton or Beck depression scores, phase of menstrual cycle, or baseline estradiol levels. However, within the patient group, peak delta prolactin responses were negatively correlated with baseline cortisol levels ($r = -.40$, $P \leq .05$) and with weekly binge frequency during the month prior to admission ($r = -.37$, $P \leq .06$). Whether this alteration in 5-HT functioning represents a causative factor in the pathogenesis of bulimia nervosa, is a sequela of the disorder, or a combination of both, cannot be stated with certainty. Prolonged dieting, which can cause 5-HT receptor

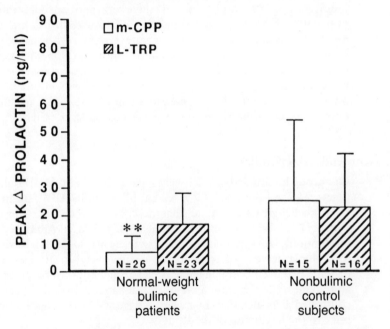

**Figure 7-3.** Mean peak delta prolactin responses (± SD) to *m*-chlorophenylpiperazine (m-CPP) and L-tryptophan (L-TRP) in normal-weight bulimic patients and in nonbulimic control subjects. (**) indicates $P < .002$, Mann-Whitney U test, bulimic patients vs. control subjects after m-CPP administration.

upregulation (Goodwin et al. 1987a, 1987b, 1987c), is an unlikely cause of the observed changes, because 20 of the 26 patients were studied after a 3- to 4-week period of relative weight stability and abstinence from binge eating and vomiting. Furthermore, the six outpatients studied had peak delta prolactin responses equivalent to the 20 abstinent inpatients. Binge eating and vomiting, which may affect central 5-HT synthesis (Kaye et al. 1988a), could result in downregulation of postsynaptic 5-HT receptors and blunted neuroendocrine responses. Although the 3- to 4-week abstinence period would be expected to allow for a return to baseline conditions, longer-term alterations in 5-HT receptor functioning as a result of

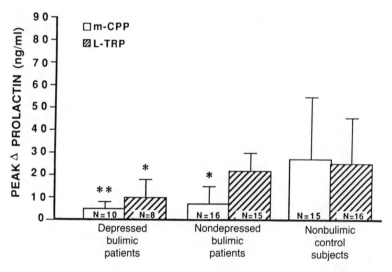

**Figure 7-4.** Mean peak delta prolactin responses ($\pm$ SD) to *m*-chlorophenylpiperazine (m-CPP) and L-tryptophan (L-TRP) in normal-weight bulimic patients with and without concurrent DSM-III major depression and in nonbulimic control subjects. (\*\*) indicates $P < .002$, Mann-Whitney U test, depressed bulimic patients vs. control subjects after m-CPP administration; (\*) indicates $P < .02$, Mann-Whitney U test, nondepressed bulimic patients vs. control subjects after m-CPP and depressed bulimic patients vs. control subjects after L-TRP. In addition, depressed bulimic patients had significantly lower peak delta prolactin responses to L-TRP than nondepressed bulimic patients ($P < .005$, unpaired Student's *t* test).

chronic binge eating and vomiting are possible. Alternatively, it may be that the blunted responses reflect a premorbid state that predisposes these patients toward bulimia nervosa and perhaps affective illness. The trend for an inverse correlation between weekly binge frequency during the month prior to admission and peak delta prolactin is compatible with either interpretation.

The mean peak delta prolactin response following L-tryptophan administration in 23 bulimic patients ($17.6 \pm 10.1$ ng/ml) was not significantly lower than that of 16 age-matched control subjects ($25.4 \pm 21.0$ ng/ml, $P = .21$, unpaired Student's t test) (Figure 7-3). However, eight bulimic patients with concurrent major depression had significantly lower peak delta prolactin levels ($9.9 \pm 7.4$ ng/ml) than did the remaining 15 bulimic patients without major depression ($21.7 \pm 8.9$ ng/ml, $P < .005$, unpaired Student's t test) (Figure 7-4). Depressed bulimic patients also had significantly lower peak delta prolactin levels than control subjects ($P < .02$, Mann-Whitney U Test).

Why do blunted responses occur in both the depressed and the nondepressed bulimic patients following m-CPP administration and only in the depressed bulimic patients following L-tryptophan? The answer probably relates to disparate mechanisms of action between these two agents. Prolactin stimulation following L-tryptophan administration involves *both* presynaptic and postsynaptic serotonergic neurons, whereas prolactin stimulation following m-CPP administration only involves postsynaptic serotonergic neurons. It is conceivable that major depression involves disruption of presynaptic serotonergic mechanisms in a state-dependent fashion. Findings from studies of decreased platelet 5-HT uptake and platelet $^3$H-labeled imipramine binding are compatible with this hypothesis (Healy and Leonard 1987). Bulimia nervosa, which is associated with a predisposition to mood disturbances, apparently involves a preferential disturbance in postsynaptic serotonergic sensitivity that may worsen with increasing cortisol production. It may be that an adaptive increase in presynaptic 5-HT output occurs in response to decreased postsynaptic serotonergic sensitivity. In effect, the decrease in postsynaptic sensitivity may be countered by an increase in presynaptic output, resulting in normal prolactin responses to -tryptophan in the nondepressed bulimic patients. This adaptive countermeasure may fail during the development of major depression. Such an hypothesis is supported by the report by Goldbloom and colleagues (1988) of increased $V_{max}$ of platelet 5-HT uptake in nondepressed bulimic patients. The results of these studies provide important new evidence for serotonergic dysregulation in normal-weight patients with bulimia nervosa and further emphasize the relevance of testing 5-HT–active drugs for the treatment of bulimia.

## Pharmacologic Studies

Pharmacologic studies in bulimia nervosa are reviewed elsewhere (Agras and McCann 1987; Garfinkel and Garner 1987; Gwirtsman et al. 1984). As in anorexia nervosa, the agents found to be effective in bulimia nervosa have major effects on 5-HT function. Imipramine (Agras et al. 1987; Pope et al. 1983), desipramine (Barlow et al. 1988; Hughes et al. 1986), phenelzine (Walsh et al. 1984), fenfluramine (Russell 1985), and fluoxetine (Freeman et al. 1988) have been found in double-blind, placebo-controlled studies to be clinically effective in a substantial proportion of patients with bulimia nervosa. Lithium (Hsu 1984) has been reported in an open trial to decrease binge eating and vomiting in normal-weight bulimic patients. In addition, acute fenfluramine administration has been shown to inhibit binge eating and vomiting in bulimia nervosa patients studied in a controlled environment (Robinson et al. 1985). The long-term usefulness of the above serotonergic medications in bulimia nervosa patients requires further study, although initial reports with antidepressant drugs appear to be favorable (Pope et al. 1985). Apart from their effects on other neurochemical systems, tricyclic antidepressants, MAO inhibitors (MAOIs), fluoxetine, and lithium are known to potentiate central 5-HT function (Blier et al. 1987; Charney et al. 1981, 1984; de Montigny and Aghajanian 1978; Fuxe et al. 1983; Wilner 1985). Preliminary data suggest that 8 weeks of fenfluramine treatment also enhances prolactin responses to intravenous L-tryptophan in humans (T. D. Brewerton, unpublished data). The effectiveness of amitriptyline (Mitchell and Groat 1984) and mianserin (Sabine et al. 1983) in the treatment of bulimia nervosa was not significantly different from placebo. However, both of these studies had unusually high placebo response rates, used fixed doses of medication, and did not report plasma drug levels. Interestingly, unlike tricyclic antidepressants and MAOIs, mianserin treatment failed to enhance the prolactin response to L-tryptophan in a group of depressed patients (Cowen 1988). Cole and Lapierre (1986) reported a case of normal-weight bulimia responsive to 1 g of orally administered L-tryptophan. However, Krahn and Mitchell (1985) reported that the antibulimic effect of L-tryptophan (1 g orally three times a day) was not significantly different from placebo in 13 normal-weight patients studied as outpatients over a 6-week period. However, in this study there is no mention of the temporal relationships among binge eating, vomiting, and L-tryptophan administration. It is imperative that L-tryptophan be administered on an empty stomach in order to avoid competition with other neutral amino acids and to promote transport into the CNS. In addition, the dose of L-tryptophan used in this study may have been too low, given that some bulimic patients may have an impaired ability to synthesize 5-HT from peripheral L-tryptophan (Kaye et al. 1988a). Higher doses of L-tryptophan have been used successfully in other psychiatric patients without adverse effects (Brewerton and Reus 1983).

## CONCLUSIONS

On theoretical grounds 5-HT is likely to play an important role in the pathophysiology of anorexia nervosa and bulimia nervosa. This chapter has presented an overview of the available data supporting this notion. The preponderance of experimental and pharmacologic evidence generally supports the hypothesis that disturbances in 5-HT function occur in the eating disorders. Decreases in levels of plasma L-tryptophan, urinary 5-HIAA, platelet imipramine binding, and basal CSF 5-HIAA in anorexia nervosa normalize upon weight restoration and appear to be starvation effects. These alterations in 5-HT function may, however, perpetuate the symptomatology of anorexia nervosa once the illness is set in motion. Some drugs that in part affect 5-HT function facilitate weight gain in conjunction with an integrated psychotherapeutic and behavioral program. Patients with bulimia nervosa, regardless of the presence of anorexia nervosa or major depression, who have been relatively weight-stable and free of binge and/or vomit episodes for at least 3 weeks have significantly blunted prolactin responses to the 5-HT agonist m-CPP. These findings indicate that postsynaptic responsivity in hypothalamo-pituitary serotonergic pathways is blunted in bulimia. Similar alterations in other serotonergic pathways at or above the level of the hypothalamus may contribute to binge eating and other behavioral symptoms in bulimic patients. The clinical response to several psychotropic agents known to potentiate 5-HT transmission further substantiates a serotonergic dysregulation hypothesis of bulimia nervosa.

## REFERENCES

Abraham SF, Beumont PJV: How patients describe bulimia or binge eating. Psychol Med 12:625–635, 1982

Agras WS, McCann U: The efficacy and role of antidepressants in the treatment of bulimia nervosa. Ann Behav Med 9:18–22, 1987

Agras WS, Dorian B, Kirkley BG, et al: Imipramine in the treatment of bulimia: a double-blind controlled study. Int J Eating Disord 6:29–38, 1987

American Psychiatric Association: Diagnostic and Statistical Manual of Mental Disorders, 3rd Edition. Washington, DC, American Psychiatric Association, 1980

American Psychiatric Association: Diagnostic and Statistical Manual of Mental Disorders, 3rd Edition, Revised. Washington, DC, American Psychiatric Association, 1987

Åsberg M, Thorén P, Träskman L: "Serotonin depression"—a biochemical subgroup within the affective disorders? Science 191:478, 1976a

Åsberg M, Träskman L, Thorïn P: 5-HIAA in the cerebrospinal fluid: a biochemical suicide predictor? Arch Gen Psychiatry 33:1193–1197, 1976b

Åsberg M, Schalling D, Träskman-Bendz L, et al: Psychobiology of suicide, impulsivity, and related phenomena, in Psychopharmacology: The Third Generation of Progress. Edited by Meltzer HY. New York, Raven Press, 1987, pp 655–668

Ballenger J, Goodwin F, Major L, et al: Alcohol and central serotonin metabolism in man. Arch Gen Psychiatry 36:224–227, 1979

Barlow J, Blouin J, Blouin A, et al: Treatment of bulimia with desipramine: a double-blind crossover study. Can J Psychiatry 33:129–133, 1988

Beumont PJV, George GCW, Pimstone BL, et al: Body weight and the pituitary response to hypothalamic releasing hormones in patients with anorexia nervosa. J Clin Endocrinol Metab 43:487–496, 1976a

Beumont PJV, George GCW, Smart DE: "Dieters" and "vomiters and purgers" in anorexia nervosa. Psychol Med 6:617–622, 1976b

Biederman J, Herzog DB, Rivinus TN, et al: Amitriptyline in the treatment of anorexia nervosa. J Clin Psychopharmacol 5:10–16, 1984a

Biederman J, Rivinus TM, Herzog DB, et al: Platelet MAO activity in anorexia nervosa patients with and without a major depressive disorder. Am J Psychiatry 141:1244–1247, 1984b

Bjerkenstedt L, Edman G, Hagenfeldt L, et al: Plasma amino acids in relation to cerebrospinal fluid monoamine metabolites in schizophrenic patients and healthy controls. Br J Psychiatry 147:276–282, 1985

Blackshear MA, Friedman RL, Sanders-Bush E: Acute and chronic effects of serotonin (5-HT) antagonists on serotonin binding sites. Naunyn Schmeidebergs Arch Pharmacol 324:125–129, 1983

Blackshear MA, Martin LL, Sanders-Bush E: Adaptive changes in the 5-HT$_2$

binding site after chronic administration of agonists and antagonists. Neuropharmacology 25:1267–1271, 1986

Blier P, de Montigny C, Chaput Y: Modifications of the serotonin system by antidepressant treatments: implications for the therapeutic response in major depression. J Clin Psychopharmacol 7:24S–35S, 1987

Blundell JE: Is there a role for serotonin (5-hydroxy-tryptamine) in feeding? Int J Obes 1:15–42, 1977

Blundell JE: Serotonin and appetite. Neuropharmacology 23:1537–1551, 1984

Blundell JE: Serotonin manipulations and the structure of feeding behaviour. Appetite 7 (suppl):39–56, 1986

Borg S, Kvande H, Liljeberg P, et al: 5-Hydroxyindoleacetic acid in cerebrospinal fluid in alcoholic patients under different clinical conditions. Alcohol 2:415–418, 1985

Boskind-Lodahl M: Cinderella's stepsisters: a feminist perspective on anorexia nervosa and bulimia. Journal of Women in Culture and Society 2:342–356, 1976

Brewerton TD, Reus VI: Lithium carbonate and L-tryptophan in the treatment of bipolar and schizoaffective disorders. Am J Psychiatry 140:757–760, 1983

Brewerton TD, Heffernan MM, Rosenthal NE: Psychiatric aspects of the relationship between eating and mood. Nutr Rev 44 (suppl):78–88, 1986a

Brewerton TD, Mueller EA, George DT, et al: Blunted prolactin response to the serotonin agonist m-chlorophenylpiperazine (m-CPP) in bulimia. Abstract of paper presented at the 15th Collegium Internationale Neuro-Psychopharmacologium Congress, San Juan, Puerto Rico, December 1986b, p 186

Brewerton TD, Mueller EA, Brandt HA, et al: Evidence for serotonin dysregulation in anorexia, in New Research: Program & Abstracts of the 140th Annual Meeting of the American Psychiatric Association, Chicago, IL, May 1987. Washington, DC, American Psychiatric Association, 1987a, p 123

Brewerton TD, Mueller EA, Brandt HA, et al: Neuroendocrine studies with 5-HT agents in bulimia, in CME Syllabus & Proceedings Summary of the 140th Annual Meeting of the American Psychiatric Association, Chicago, IL, May 1987. Washington, DC, American Psychiatric Association, 1987b, pp 85–86

Brewerton TD, Murphy D, Mueller EA, et al: The induction of migraine-like headaches by the serotonin agonist, *m*-chlorophenylpiperazine. Clin Pharmacol Ther 43:605–609, 1988

Brown GL, Ebert MH, Goyer PF, et al: Aggression, suicide, and serotonin: relationships to CSF amine metabolites. Am J Psychiatry 139:741–746, 1982

Bulik CM: Drug and alcohol abuse by bulimic women and their families. Am J Psychiatry 144:1604–1606, 1987

Carli M, Samanin R: Potential anxiolytic properties of 8-hydroxy-2-(di-N-propylamino)tetralin, a selective serotonin$_{1A}$ receptor agonist. Psychopharmacology (Berlin) 94:84–91, 1988

Casper RC: On the emergence of bulimia nervosa as a syndrome: a historical view. Int J Eating Disord 2:3–16, 1983

Casper RC, Eckert ED, Halmi DA, et al: Bulimia: its incidence and clinical importance in patients with anorexia nervosa. Arch Gen Psychiatry 37:1030–1035, 1980

Chakrabarty AS, Pillai RV, Anand BK, et al: Effect of cyproheptadine on the electrical activity of the hypothalamic feeding centres. Brain Res 6:561–569, 1967

Charney DS, Menkes DB, Heninger GR: Receptor sensitivity and the mechanism of action of antidepressant treatment. Arch Gen Psychiatry 38:1160–1180, 1981

Charney DS, Heninger GR, Sternberg DE: Serotonin function and mechanism of action of antidepressant treatment effects of amitriptyline and desipramine. Arch Gen Psychiatry 41:359–365, 1984

Charney DS, Woods SW, Goodman WK, et al: Serotonin function in anxiety. II. Effects of the serotonin agonist MCPP in panic disorder patients and healthy subjects. Psychopharmacology (Berlin) 92:14–24, 1987

Cole W, Lapierre YD: The use of tryptophan in normal-weight bulimia. Can J Psychiatry 31:755–756, 1986

Cooper PJ, Fairburn CG: The depressive symptoms of bulimia nervosa. Br J Psychiatry 148:268–274, 1986

Coppen AJ, Gupta RK, Eccleston EG, et al: Plasma tryptophan in anorexia nervosa. Lancet 1:961, 1976

Cowen PJ: Prolactin response to tryptophan during mianserin treatment. Am J Psychiatry 145:740–741, 1988

Cowen PJ, Charig EM: Neuroendocrine responses to intravenous tryptophan in major depression. Arch Gen Psychiatry 44:958–966, 1987

Crisp AH: Anorexia nervosa at a normal weight! The abnormal weight control syndrome. Int J Psychiatry Med 11:203–234, 1981

Crisp AH, Hsu LKG, Harding B: The starving hoarder and voracious spender: stealing in anorexia nervosa. J Psychosom Res 24:225–231, 1980

Davis DD, Dunlop SR, Shea P, et al: Biological stress responses in high and low trait anxious students. Biol Psychiatry 20:843–851, 1985

de Montigny C, Aghajanian GK: Tricyclic antidepressants: long-term treatment increases responsivity of rat forebrain neurons to serotonin. Science 202:1303–1306, 1978

Dourish CT, Hutson PH, Kennett GA, et al: 8-OH-DPAT-induced hyperphagia: its neural basis and possible therapeutic relevance. Appetite 7 (suppl):127–140, 1986

Dunner DL, Goodwin FK: Effect of L-tryptophan on brain serotonin metabolism in depressed patients. Arch Gen Psychiatry 26:364–366, 1972

Eccleston D, Ashcroft GW, Crawford TBB, et al: Effects of tryptophan administration on 5-HIAA in cerebrospinal fluid in man. J Neurol Neurosurg Psychiatry 33:269–272, 1970

Endicott J, Spitzer RL: A diagnostic interview: the Schedule for Affective Disorders and Schizophrenia. Arch Gen Psychiatry 35:837–844, 1978

Engel JA, Hjorth S, Svensson K, et al: Anticonflict effect of the putative serotonin receptor agonist 8-hydroxy-2-(DI-n-propylamino)tetralin (8-OH-DPAT). Eur J Pharmacol 105:365–368, 1984

Ferguson JM: Bulimia: a potentially fatal syndrome. Psychosomatics 26:252–253, 1985

Freeman CPL, Morris JE, Cheshire KE, et al: A double-blind controlled trial of fluoxetine versus placebo for bulimia nervosa. Abstract of paper presented at the Third International Conference on Eating Disorders, New York, April 1988

French AP, Nelson HL: Genital self-mutilation in women. Arch Gen Psychiatry 27:618–620, 1972

Fuxe K, Ogren S-O, Agnati LF, et al: Chronic antidepressant treatment and central 5-HT synapses. Neuropharmacology 22:389–400, 1983

Gandolfi O, Barbaccia ML, Costa E: Different effects of serotonin antagonists on $^3$H-mianserin and $^3$-H-ketanserin recognition sites. Life Sci 36:713–721, 1985

Garattini S, Bizzi A, Caccia S, et al: Progress in assessing the role of serotonin in the control of food intake. Clin Neuropharmacol 11 (suppl 1):S8–S32, 1988

Gardner CR: Pharmacological studies on the role of serotonin in animal models of anxiety, in Neuropharmacology of Serotonin. Edited by Green AR. Oxford, UK, Oxford University Press, 1985, pp 281–325

Garfinkel PE, Garner DM: Subtypes of anorexia nervosa, in Anorexia Nervosa: A Multidimensional Perspective. Edited by Garfinkel PE, Garner DM. New York, Brunner/Mazel, 1982, pp 40–57

Garfinkel PE, Garner DM: The Role of Drug Treatments for Eating Disorders. New York, Brunner/Mazel, 1987

Garfinkel PE, Moldofsky H, Garner DM: The heterogeneity of anorexia nervosa: bulimia as a distinct subgroup. Arch Gen Psychiatry 37:1036–1040, 1980

George DT, Brewerton TD, Jimerson DC: Comparison of lactate-induced anxiety in bulimic patients and healthy controls. Psychiatry Res 21:213–220, 1987

Gerner RH, Cohen DJ, Fairbanks L, et al: CSF neurochemistry of women with anorexia nervosa and normal women. Am J Psychiatry 141:948–949, 1984

Gill K, Amit Z: Effects of serotonin uptake blockade on food, water, and ethanol consumption in rats. Alcoholism (NY 11:444–449, 1987

Gillberg C: Low dopamine and serotonin levels in anorexia nervosa. Am J Psychiatry 140:948–949, 1983

Gjerris A, Sorensen AS, Rafaelsen OJ, et al: 5-HT and 5-HIAA in cerebrospinal fluid in depression. J Affective Disord 12:13–22, 1987

Glaser T, Traber J: Binding of the putative anxiolytic TVX Q 7821 to hippocampal 5-hydroxytryptamine (5-HT) recognition sites. Naunyn Schmeidebergs Arch Pharmacol 329:211–215, 1985

Glue PW, Cowen PJ, Nutt DJ, et al: The effect of lithium on 5-HT mediated neuroendocrine responses and platelet 5-HT receptors. Psychopharmacology (Berlin) 90:398–402, 1986

Goldbloom DS: Serotonin in eating disorders: theory and therapy, in The Role of Drug Treatments for Eating Disorders. Edited by Garfinkel PE, Garner DM. New York, Brunner/Mazel, 1987, pp 124–149

Goldbloom DS, Hicks LK, Garfinkel PE: Platelet serotonin uptake in bulimia nervosa, in New Research: Program and Abstracts of the 141st Annual Meeting of the American Psychiatric Association, Montreal, Quebec,

May 1988. Washington, DC, American Psychiatric Association, 1988, p 137

Goldney RD, Simpson IG: Female genital self-mutilation, dysorexia and the hysterical personality: the Caenis syndrome. Can Psychiatr Assoc J 20:435–441, 1975

Goodwin GM, Fairburn CG, Cowen PJ: Dieting changes serotonergic function in women, not men: implications for the aetiology of anorexia nervosa. Psychol Med 17:839–842, 1987a

Goodwin GM, Fairburn CG, Cowen PJ: The effects of dieting and weight loss on neuroendocrine responses to tryptophan, clonidine and apomorphine in volunteers: important implications for neuroendocrine investigations in depression. Arch Gen Psychiatry 44:952–957, 1987b

Goodwin GM, Fraser S, Stump K, et al: Dieting and weight loss in volunteers increases the number of alpha2-adrenoceptors and 5-HT receptors on blood platelets without effect on [$^3$H]imipramine binding. J Affective Disord 12:267–274, 1987c

Gross HA, Ebert MH, Faden VB, et al: A double-blind controlled trial of lithium carbonate in primary anorexia nervosa. J Clin Psychopharmacol 1:376–381, 1984

Gwirtsman HE, Kaye W, Weintraub M, et al: Pharmacologic treatment of eating disorders. Psychiatr Clin North Am 7:863–877, 1984

Hagenfeldt L, Bjerkenstedt L, Edman G, et al: Amino acids in plasma and CSF and monomine metabolites in CSF: interrelationship in healthy subjects. J Neurochem 42:833–837, 1984

Halmi KA, Eckert E, LaDu TJ, et al: Anorexia nervosa: treatment efficacy of cyproheptadine and amitriptyline. Arch Gen Psychiatry 43:177–181, 1986

Harris RT: Bulimarexia and related serious eating disorders with medical complications. Ann Intern Med 99:800–807, 1983

Hatsukami DK, Mitchell JE, Eckert ED: Eating disorders: a variant of mood disorders? Psychiatr Clin North Am 7:349–365, 1984

Healy D, Leonard BE: Monoamine transport in depression: kinetics and dynamics. J Affective Disord 12:91–103, 1987

Hsu LKG: Treatment of bulimia with lithium. Am J Psychiatry 141:1260–1262, 1984

Hsu LKG, Crisp AH, Harding B: Outcome of anorexia nervosa. Lancet 1:61–65, 1979

Hudson JI, Pope HG, Jonas JM, et al: Phenomenologic relationship of eating disorders to major affective disorder. Psychiatry Res 9:345–354, 1984

Hudson JI, Pope HG, Yurgelun-Todd D, et al: A controlled study of lifetime prevalence of affective and other psychiatric disorders in bulimic outpatients. Am J Psychiatry 144:1283–1287, 1987

Hughes PL, Wells LA, Cunningham CJ, et al: Treating bulimia with desipramine: a double-blind, placebo-controlled study. Arch Gen Psychiatry 43:182–186, 1986

Jacobs MB, Schneider JA: Medical complications of bulimia: a prospective evaluation. Q J Med 54:177–182, 1985

Johnson C, Larson R: Bulimia: an analysis of moods and behavior. Psychosom Med 44:341–351, 1982

Johnston AL, File SE: 5-HT and anxiety: promises and pitfalls. Pharmacol Biochem Behav 24:1467–1470, 1986

Johnston JL, Leiter LA, Burrow GN, et al: Excretion of urinary catecholamine metabolites in anorexia nervosa: effect of body composition and energy intake. Am J Clin Nutr 40:1001–1006, 1984

Jones DJ, Fox MM, Babigian HM, et al: Epidemiology of anorexia nervosa in Monroe County, New York: 1960–1976. Psychosom Med 42:551–558, 1980

Kaye WH, Ebert MH, Gwirtsman HE, et al: Differences in brain serotonergic metabolism between nonbulimic and bulimic patients with anorexia nervosa. Am J Psychiatry 141:1598–1601, 1984a

Kaye WH, Ebert MH, Raleigh M, et al: Abnormalities in CNS monoamine metabolism in anorexia nervosa. Arch Gen Psychiatry 41:350–355, 1984b

Kaye WH, Gwirtsman HE, Brewerton TD, et al: Bingeing behavior and plasma amino acids: a possible involvement of brain serotonin in bulimia. Psychiatry Res 23:31–43, 1988a

Kaye WH, Gwirtsman HE, George DT, et al: CSF 5-HIAA concentrations in anorexia nervosa: reduced values in underweight subjects normalize after weight gain. Biol Psychiatry 23:102–105, 1988b

Kendell RE, Hall DJ, Hailey A, et al: The epidemiology of anorexia nervosa. Psychol Med 3:200–203, 1973

Kiriike N, Nishiwaki S, Izumiya Y, et al: Thyrotropin, prolactin, and growth hormone responses to thyrotropin-releasing hormone in anorexia nervosa and bulimia. Biol Psychiatry 22:167–176, 1987

Knapp S, Mandell AJ: Short- and long-term lithium administration: effects on the brain's serotonergic biosynthetic systems. Science 180:645–647, 1973

Kornhaber AK: The stuffing syndrome. Psychosomatics 11:580–584, 1970

Koskiniemi M, Laakso J, Kuurne T, et al: Indole levels in human lumbar and ventricular cerebrospinal fluid and the effect on L-tryptophan administration. Acta Neurol Scand 71:127–132, 1985

Krahn D, Mitchell J: Use of L-tryptophan in treating bulimia. Am J Psychiatry 142:1130, 1985

Lacey JH, Crisp AH: Hunger, food intake and weight: the impact of clomipramine on a refeeding anorexia nervosa population. Postgrad Med J 56 (suppl 1):79–85, 1980

Leibowitz SF, Shor-Posner G: Brain serotonin and eating behavior. Appetite 7 (suppl):1–14, 1986

Li ETS, Anderson GH: 5-Hydroxytryptamine: a modulator of food composition but not quantity? Life Sci 34:2453–2460, 1984

Linnoila M, Virkkunen M, Scheinin M, et al: Low cerebrospinal fluid 5-hydroxyindoleacetic acid concentration differentiates impulsive from nonimpulsive violent behavior. Life Sci 33:2609–2614, 1983

Loro AD, Orleans CS: Binge eating in obesity: preliminary findings and guidelines for behavioral analysis and treatment. Addict Behav 6:155–166, 1981

Lydiard RB, Brady KT, O'Neil PM, et al: Precursor amino acid concentrations in normal weight bulimics and normal controls. Prog Neuropsychopharmacol Biol Psychiatry 12:893–898, 1988

Macaron C, Wilber JF, Green O, et al: Studies of growth hormone (GH), thyrotropin (TSH) and prolactin (PRL) secretion in anorexia nervosa. Psychoneuroendocrinology 3:181–185, 1978

Marcus MD, Wing RR: Binge eating among the obese. Ann Behav Med 9:23–27, 1987

Mitchell JE, Groat R: A placebo-controlled, double-blind trial of amitriptyline in bulimia. J Clin Psychopharmacol 4:186–193, 1984

Mitchell JE, Pyle RL, Eckert ED, et al: Electrolyte and other physiological abnormalities in patients with bulimia. Psychol Med 13:272–278, 1983

Mitchell JE, Hatsukami D, Eckert ED, et al: Characteristics of 275 patients with bulimia. Am J Psychiatry 142:482–485, 1985

Mitchell JE, Hatsukami K, Pyle RL, et al: The bulimia syndrome: course of the illness and associated problems. Compr Psychiatry 27:165–170, 1986

Morgan HG, Russell GFM: Value of family background and clinical features as predictors of long-term outcome in anorexia nervosa: four-year follow-up study of 41 patients. Psychol Med 5:355–371, 1975

Mueller EA, Murphy DL, Sunderland T: Neuroendocrine effects of *m*-chlorophenylpiperazine, a serotonin agonist, in humans. J Clin Endocrinol Metab 61:1179–1184, 1985

Mueller EA, Murphy DL, Sunderland T: Further studies of the putative serotonin agonist, *m*-chlorophenylpiperazine: evidence for a serotonin receptor mediated mechanism of action in humans. Psychopharmacology (Berlin) 89:388–391, 1986

Murphy DL, Mueller EA, Garrick NA, et al: Use of serotonergic agents in the clinical assessment of central serotonin function. J Clin Psychiatry 47 (suppl):9–15, 1986

Murphy JM, Walker MB, Gatto GJ, et al: Monoamine uptake inhibitors attenuate ethanol intake in alcohol-preferring (P) rats. Alcohol 2:349–352, 1985

Myers RD, Melchior CL: Alcohol and alcoholism: role of serotonin, in Serotonin in Health and Disease, Vol 2. Edited by Essman WB. New York, Spectrum Publications, 1978, pp 373–430

Naranjo CA, Sellers EM, Roach CA, et al: Zimelidine-induced variations in alcohol intake by nondepressed heavy drinkers. Clin Pharmacol Ther 35:374–381, 1984

Naranjo CA, Sellers EM, Lawrin MO: Modulation of ethanol intake by serotonin uptake inhibitors. J Clin Psychiatry 47 (suppl):16–22, 1986

Naranjo CA, Sellers EM, Sullivan JT, et al: The serotonin uptake inhibitor citalopram attenuates ethanol intake. Clin Pharmacol Ther 41:266–274, 1987

Oomura Y, Ono T, Sugimori M, et al: Effects of cyproheptadine on the feeding and satiety centers in the rat. Pharmacol Biochem Behav 1:449–459, 1973

Palmer RL: The dietary chaos syndrome: a useful new term? Br J Med Psychol 52:43–52, 1979

Peroutka SJ: Selective interaction of novel anxiolytics with 5-hydroxytryptamine$_{1A}$ receptors. Biol Psychiatry 20:971–979, 1985

Pope HG, Hudson JI, Jonas IM, et al: Bulimia treated with imipramine: a placebo-controlled, double-blind study. Am J Psychiatry 140:554–558, 1983

Pope HG, Hudson JI, Jonas IM, et al: Antidepressant treatment of bulimia: a two-year follow-up study. J Clin Psychopharmacol 5:320–327, 1985

Riederer P, Toifl K, Kruzik P: Excretion of biogenic amine metabolites in anorexia nervosa. Clin Chim Acta 123:27–32, 1982

Robinson PH, Checkley SA, Russell GFM: Suppression of eating by fenfluramine in patients with bulimia nervosa. Br J Psychiatry 146:169–176, 1985

Rogers PJ, Blundell JE: Effect of anorexic drugs on food intake and the microstructure of eating in human subjects. Psychopharmacology (Berlin) 66:159–165, 1979

Rosenthal RJ, Rinzler C, Wallsh R, et al: Wrist-cutting syndrome: the meaning of a gesture. Am J Psychiatry 128:1363–1368, 1972

Rosenzweig M, Spruill J: Twenty years after Twiggy: a retrospective investigation of bulimic-like behaviors. Int J Eating Disord 6:59–65, 1987

Rothenberg A: Eating disorder as a modern obsessive-compulsive syndrome. Psychiatry 49:45–53, 1986

Roy A, Virkkunen M, Linnoila M: Reduced central serotonin turnover in a subgroup of alcoholics? Prog Neuropsychopharmacol Biol Psychiatry 11:173–177, 1987

Russell GFM: The nutritional disorder in anorexia nervosa. J Psychosom Res 11:141–149, 1967

Russell GFM: Bulimia nervosa: an ominous variant of anorexia nervosa. Psychol Med 9:429–448, 1979

Russell GFM: Bulimia revisited. Int J Eating Disord 4:681–692, 1985

Sabine EJ, Yonace A, Farrington AJ, et al: Bulimia nervosa: a placebo controlled double-blind therapeutic trial of mianserin. Br J Clin Pharmacol 15:195S–202S, 1983

Samanin R, Mennini T, Ferraris A: m-Chlorophenylpiperazine: a central serotonin agonist causing powerful anorexia in rats. Naunyn Schmiedebergs Arch Pharmacol 308:159–163, 1982

Scott DW: Alcohol and food abuse: some comparisons. Br J Addict 78:339–349, 1984

Silverstone T, Goodall E: Serotonergic mechanisms in human feeding: the pharmacological evidence. Appetite 7 (suppl):85–97, 1986

Simpson MA: The phenomenology of self-mutilation in a general hospital setting. Can Psychiatr Assoc J 20:429–434, 1975

Society of Actuaries and Association of Life Insurance Medical Directors of America: Build Study, 1979. Chicago, IL, Society of Actuaries, 1980

Strober M: The significance of bulimia in juvenile anorexia nervosa: an exploration of possible etiologic factors. Int J Eating Disord 1:28–43, 1981

Stunkard AJ: Eating patterns and obesity. Psychiatr Q 33:284–292, 1959

Swift WJ, Ritholz M, Kalin NH, et al: A follow-up study of thirty hospitalized bulimics. Psychosom Med 49:45–55, 1987

Szmukler GI, McCance C, McCrone L, et al: Anorexia nervosa: a psychiatric case register study from Aberdeen. Psychol Med 16:49–58, 1986

Theander S: Anorexia nervosa: a psychiatric investigation of 94 female patients. Acta Psychiatr Scand 214 (suppl):7–194, 1970

Theander S: Outcome and prognosis in anorexia nervosa and bulimia: some results of previous investigations, compared with those of a Swedish long-term study. J Psychiatr Res 19:493–508, 1985

Valzelli L, Bernasconi S, Dalessandro M: Effect of tryptophan administration on spontaneous and p-PCA-induced muricidal aggression in laboratory rats. Pharmacol Res Commun 13:891–897, 1981

Vigersky RA, Loriaux DL: The effect of cyproheptadine in anorexia nervosa: a double-blind trial, in Anorexia Nervosa. Edited by Vigersky RA. New York, Raven Press, 1977, pp 349–356

Vigersky RA, Loriaux DL, Andersen AE, et al: Delayed pituitary hormone response to LRF and TRF in patients with anorexia nervosa and with secondary amenorrhea associated with simple weight loss. J Clin Endocrinol Metab 43:893–900, 1976

Waldhauser F, Toifl K, Spona J, et al: Diminished prolactin response to thyrotropin and insulin in anorexia nervosa. J Clin Endocrinol Metab 59:538–541, 1984

Walsh BT, Stewart JW, Roose SP, et al: Treatment of bulimia with phenelzine: a double-blind, placebo-controlled study. Arch Gen Psychiatry 41:1105–1109, 1984

Walsh BT, Roose SP, Glassman AH, et al: Bulimia and depression. Psychosom Med 47:123–131, 1985

Weiss SR, Ebert MH: Psychological and behavioral characteristics of normal-weight bulimics and normal-weight controls. Psychosom Med 45:293–303, 1983

Weizman R, Carmi M, Tyano S, et al: High affinity [$^3$H]imipramine binding and serotonin uptake to platelets of adolescent females suffering from anorexia nervosa. Life Sci 38:1235–1242, 1986

Wilner P: Antidepressants and serotonergic neurotransmission: an integrative review. Psychopharmacology (Berlin) 85:387–404,1985

Wurtman JJ, Wurtman RJ, Growdon JH, et al: Carbohydrate craving in obese people: suppression by treatments affecting serotonergic transmission. Int J Eating Disord 1:2–15, 1981

Wurtman RJ, Wurtman JJ: Carbohydrate craving, obesity and brain serotonin. Appetite 7 (suppl):99–103, 1986

Young WF, Laws ER, Sharbrough FW, et al: Human monoamine oxidase: lack of brain and platelet correlation. Arch Gen Psychiatry 43:604–609, 1986

Zemishlany Z, Modai I, Apter A, et al: Serotonin (5-HT) uptake by blood platelets in anorexia nervosa. Acta Psychiatr Scand 75:127–130, 1987

Zohar J, Insel TR: Obsessive-compulsive disorder: psychobiological approaches to diagnosis, treatment, and pathophysiology. Biol Psychiatry 22:667–687, 1987

# Chapter 8

## Serotonin in Suicide, Violence, and Alcoholism

Alec Roy, M.B.
Matti Virkkunen, M.D.
Markku Linnoila, M.D., Ph.D.

# Chapter 8

# Serotonin in Suicide, Violence, and Alcoholism

The first study of central monoamine metabolites in patients exhibiting suicidal behaviors was carried out by Åsberg et al. (1976). They found a bimodal distribution of levels of the serotonin (5-hydroxytryptamine [5-HT]) metabolite 5-hydroxyindoleacetic acid (5-HIAA) in the lumbar cerebrospinal fluid (CSF) of 68 depressed patients. Åsberg et al. also made the observation that significantly more of the depressed patients in the "low" CSF 5-HIAA group had attempted suicide in comparison to those in the "high" CSF 5-HIAA group. This finding led to these authors' proposal that low CSF 5-HIAA levels may be associated with suicidal behavior.

Subsequently, Brown et al. (1979, 1982), in two studies of personality-disordered individuals, also found that patients with a history of suicidal behavior had significantly lower CSF 5-HIAA levels than patients without such a history. Since then, other studies in personality-disordered, schizophrenic, and depressed patients have also reported an association between low levels of CSF 5-HIAA and aggressive and suicidal behaviors, although there have also been negative reports (reviewed in Åsberg et al. 1986).

It is of note that low CSF 5-HIAA levels have been found to be particularly associated with violent suicide attempts. In fact, Träskman et al. (1981) reported that CSF 5-HIAA levels were significantly lower only among those patients who had made a violent suicide attempt (i.e., hanging, drowning, shooting, gassing, several deep cuts), and that levels were not reduced among those patients who had made a nonviolent suicide attempt (i.e., overdosage). More recently, Banki and Arato (1983) found among 141 psychiatric patients suffering from depression, schizophrenia, alcoholism, or adjustment disorder, that levels of CSF 5-HIAA were significantly lower in the violent suicide attempters in all four diagnostic categories.

## POSTMORTEM STUDIES OF SUICIDE VICTIMS

Over the years most postmortem studies of the brains of suicide victims have focused on the serotonergic system. Some, but not all, of the neurochemical studies have reported modest decreases in 5-HT itself, or in its metabolite, 5-HIAA, in either the brain stem or the frontal cortex. The few studies that have examined norepinephrine, dopamine, or the dopamine metabolite homovanillic acid (HVA) have tended to be negative (reviewed in Stanley et al. 1986). There have been few postmortem brain studies of the enzymes involved in catecholamine metabolism. No changes in monoamine oxidase (MAO) activity have been reported (reviewed in Mann et al. 1986).

Four of five postmorten brain receptor studies using $^3$H-labeled imipramine as the ligand have reported significant decreases in the presynaptic binding of this ligand to serotonergic neurons in suicide victims (reviewed in Stanley et al. 1986). Stanley and Mann (1983), using $^3$H-labeled spiroperidol as the ligand, have also reported a significant increase in postsynaptic 5-HT2 binding sites among suicide victims who used violent methods to end their lives.

Taken together, these postmortem neurochemical and receptor studies tend to support the hypothesis that diminished central 5-HT metabolism (as evidenced by reduced presynaptic imipramine binding, reduced levels of 5-HT and 5-HIAA, and upregulation of the postsynaptic 5-HT2 receptors) is associated with suicide. However, most of these studies do not compare suicide victims with depressed patients who have died for reasons other than suicide. Thus, the possibility exists that the abnormality of central 5-HT metabolism found in postmortem studies of suicide victims may apply to depressive illness in general. Also, Mann et al. (1986a, 1986b) reported a significant increase in β-adrenergic receptor binding in the frontal cortex of suicide victims; in animal studies similar changes have been observed after the subjects have been exposed to uncontrollable stress (Weis, personal communication).

## RELATIONSHIP OF CEREBROSPINAL FLUID 5-HYDROXYINDOLEACETIC ACID LEVELS TO BEHAVIOR

### Violent Offenders

The relationship of low CSF 5-HIAA levels to behavior has been further explored by Linnoila et al. (1983). They examined CSF 5-HIAA levels among 36 male murderers and attempted murderers undergoing an intensive 1- to 2-month court-ordered forensic psychiatric evaluation at the Department of Psychiatry, University of Helsinki. Impulsivity was defined by the characteristics of the index crime. Patients with a clear premeditation of their act were classified as nonimpulsive, while patients without established premeditation (attacking without provocation and not knowing the victim) were classified as impulsive. The source of this information was the police report concerning the index crime. Using this method, 9 of these 36 violent offenders were classified as nonimpulsive and 27 as impulsive.

Psychiatric diagnoses in the violent offenders were made according to DSM-III criteria by a forensic psychiatrist who was aware of the criminal records of the patients. The impulsive patients had either intermittent explosive disorder or antisocial personality disorder, whereas the nonimpulsive patients had either paranoid or passive-aggressive personality disorder. All the impulsive patients also met DSM-III criteria for alcohol abuse. The authors found that CSF 5-HIAA levels were significantly lower among the two groups of impulsive violent offenders than among the nonimpulsive violent offenders (Figure 8-1). Also, the 17 offenders (14 impulsive and 3 nonimpulsive) who had committed more than one violent crime had CSF 5-HIAA levels significantly lower than those found among offenders who had committed only one violent crime (mean $67.9 \pm 12.1$ vs. $87.1 \pm 23.7$ nm, $P < .02$). Furthermore, the violent offenders who at some time had attempted suicide were found to have significantly lower CSF 5-HIAA levels than the violent offenders who had never attempted suicide. There was no difference in CSF free testosterone concentrations between the groups (Roy et al. 1986b). This finding is of relevance because in certain animal models high testosterone concentrations have been associated with aggressiveness (Selmanoff et al. 1977).

### Arsonists

Virkkunen et al. (1987) performed lumbar punctures on 20 male arsonists also undergoing intensive forensic evaluation. Impulsive fire setters were studied because they show relatively little interpersonal aggressive behavior, thus allowing an examination of whether low CSF 5-HIAA levels are primarily associated with impulsivity or aggressivity. Impulsiveness in these patients was defined as a sudden uncontrollable urge to set a fire. Arsonists setting fires for economic gain, such as in cases of insurance fraud, were excluded from the sample.

The arsonists were matched for age, height, and sex with a subgroup of the previously studied violent offenders, and they were also compared with 10 nonarsonist control subjects. The results showed that CSF 5-HIAA levels were significantly lower among the arsonists than among the other two groups. CSF levels of the norepinephrine metabolite 3-methoxy-4-hydroxyphenylglycol (MHPG) were also significantly lower among the arsonists (Figure 8-2). When subjects who had made past violent suicide attempts were excluded from groups, the arsonists again showed significantly lower CSF 5-HIAA and CSF MHPG levels than the other two groups.

All of the arsonists also met DSM-III criteria for borderline per-

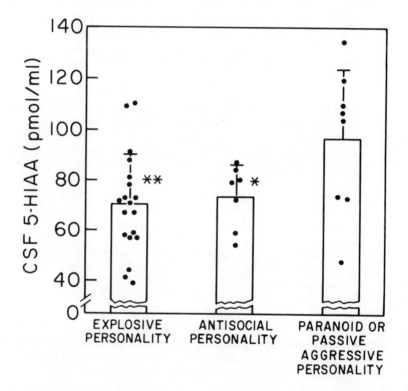

**Figure 8-1.**     CSF 5-HIAA levels in three diagnostic subgroups of violent offenders. ANOVA $P < .01$ among the groups. Double asterisk indicates $P < .01$ (by two-tailed Student's $t$ test): explosive personality levels lower than those of paranoid or passive-aggressive personality; single asterisk indicates $P < .05$ (by two-tailed Student's $t$ test): antisocial personality levels lower than those of paranoid or passive-aggressive personality. Redrawn from Linnoila et al., *Life Sciences* 33:2609–2614, 1983, with permission of Pergamon Press, Inc. Copyright 1983, Pergamon Press.

sonality disorder, and many of them had exhibited occasional explosive behavior and alcohol abuse. Only 1 of the 20 arsonists met DSM-III criteria for a major depressive episode. Virkkunen et al. also noted that among the arsonists the motive of revenge was not significantly associated with low CSF 5-HIAA levels and, furthermore, that the arsonists did not consider themselves to be aggressive. Similarly, Åsberg et al. (1987) observed that CSF 5-HIAA levels were not associated with self-reported aggressive feelings.

The results of these studies of violent offenders and arsonists have led to the hypothesis that low CSF 5-HIAA levels may be associated with poor impulse control. Although CSF 5-HIAA levels are an

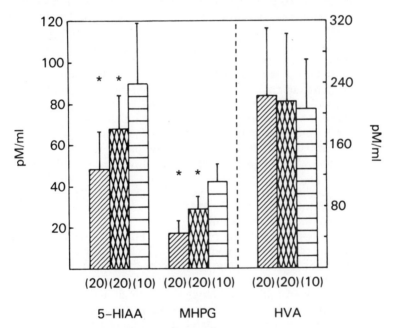

**Figure 8-2.** CSF monoamine metabolites in arsonists, violent offenders, and male control subjects. Bars representing arsonists are diagonally hatched, and bars representing violent offenders are cross-hatched. The control subjects were age- and sex-matched to the arsonists. 5-HIAA = 5-hydroxyindoleacetic acid; MHPG = 3-methoxy-4-hydroxyphenylglycol; HVA = homovanillic acid. Reproduced from Virkkunen et al., *Archives of General Psychiatry* 44:241–247, 1987, with permission of the American Medical Association. Copyright 1987, American Medical Association.

imprecise indicator of brain 5-HT turnover, we, and others, have speculated that some individuals may have a defect in their central 5-HT metabolism that manifests itself in poor impulse control, leading to either attempts at suicide or violence toward others (Roy et al. 1986a, 1986b, 1987a, 1987b, 1988c).

## Pathological Gambling

In order to further investigate possible biological substrates of impulsivity, we recently studied a relatively nonviolent group of individuals putatively with poor impulse control. Pathological gambling is classified in DSM-III as a disorder of impulse control. It is a disorder characterized by a chronic and progressive inability to resist the urge to gamble that compromises, disrupts, or damages family, personal, and vocational pursuits. This disorder is of unknown etiology (Wolkowitz et al. 1985). We compared 20 gamblers, 19 of them male, with 20 male control subjects. There were no significant differences between the groups in CSF levels of 5-HIAA. Thus, we did not confirm our initial hypothesis that pathological gambling is associated with low CSF 5-HIAA levels. However, Linnoila et al. (1983) found that the violent, impulsive criminals with the lowest CSF 5-HIAA levels were those with associated antisocial personality disorder, which in DSM-III is an exclusive criterion for pathological gambling. Thus, by following DSM-III criteria for pathological gambling, and by excluding gamblers with an antecedent antisocial personality disorder, we may have excluded gamblers who may have had low CSF 5-HIAA concentrations.

We did find that gamblers showed both a significantly higher centrally produced fraction of CSF MHPG and significantly greater urinary outputs of norepinephrine than did control subjects (Roy et al. 1988b). These findings are in accord with Zuckerman's hypothesis that the personality trait of sensation seeking may underlie risk-taking behavior, including pathological gambling, and that the biological substrate associated with sensation seeking may be an increased tonic activity of the central noradrenergic system (Zuckerman 1984; Zuckerman et al. 1983).

## Alcoholic Patients

There have been several CSF monoamine metabolite studies among alcoholic patients (reviewed in Roy and Linnoila 1989). In Japan, Takahashi et al. (1974) obtained 43 CSF specimens from 30 alcoholic patients. The authors found no significant difference in CSF 5-HIAA levels between alcoholic patients studied 1 week after alcohol withdrawal and control subjects. However, when the alcoholic patients were divided into two subgroups, according to whether or not they had withdrawal symptomatology, it was found that only the subgroup with florid withdrawal symptoms had significantly lower CSF 5-HIAA levels than control subjects. Lumbar punctures repeated in this subgroup 4 weeks later showed that the CSF 5-HIAA levels remained low. Thus, Takahashi et al. speculated that abnormality of brain 5-HT metabolism might be a factor causing both the abstinent symptoms and the depressive mood symptoms, including suicidal ideation, that are commonly found among alcoholic patients. In another study of alcoholic patients, Major et al. (1987) reported that CSF levels of the dopamine metabolite HVA were significantly reduced in patients having an alcohol-withdrawal syndrome. However, it should be noted that CSF 5-HIAA and HVA levels have shown strong positive correlations in most studies.

Ballenger et al. (1979) studied CSF monoamine metabolite levels in alcoholic patients compared with a control group of personality disorder patients who did not have significant drinking histories. The alcoholic patients were studied twice—within 48 hours of being admitted to the hospital when being detoxified and after 4 weeks of abstinence. The authors found that CSF 5-HIAA levels were significantly lower among the alcoholic patients during their abstinent phase than during the postintoxication phase, or when compared to the control subjects. In 1981, Banki, working in Hungary, reported lumbar punctures on 36 female alcoholic patients and found a significant negative correlation between the number of days abstinent and CSF 5-HIAA levels, thus confirming the earlier observation of Ballenger et al.

The interpretation of these findings suggested by Ballenger et al. was that alcoholic patients may have preexisting low brain 5-HT levels that are transiently raised by alcohol consumption. These authors theorized that this transient increase in turn may eventually lead to further depletion of brain 5-HT. They speculated that a pattern may become established whereby the alcoholic individual drinks repeatedly in order pharmacologically to modify a 5-HT deficiency in his or her brain. Interestingly, there is considerable evidence from animal studies that alcohol does indeed cause significant alterations in central indoleamine and catecholamine turnover (reviewed in Roy and Linnoila 1986). Ballenger et al. (1979) also noted that the genetic strains of rats that prefer to drink alcohol rather than water have low levels of brain 5-HT.

In another alcohol-related CSF study, Rosenthal et al. (1980) examined the CSF monoamine metabolite data obtained from 69 medication-free depressed patients. These patients were classified into subgroups according to family history. Those patients with a history of alcoholism in a first- or second-degree relative were compared with those without such a history. The depressed patients with a positive family history of alcoholism were found to have significantly lower CSF levels of 5-HIAA, and also of MHPG, than those patients without a family history of alcoholism. However, in a negative report, Lidberg et al. (1985) found that alcoholic homicidal offenders had significantly higher CSF 5-HIAA levels than nonalcoholic murderers.

Branchey et al. (1984) reported a peripheral relationship between 5-HT and aggressive behavior among alcoholic patients. Studying the ratio of tryptophan to other neutral amino acids in serum, these authors found significantly lower ratios, suggesting a deficiency in brain 5-HT, among the alcoholic patients who had been arrested for assaultive behavior when compared either to other alcoholic patients or to nonalcoholic control subjects.

## SUICIDAL BEHAVIOR AMONG ALCOHOLIC PATIENTS

In previous sections of this chapter 5-HT has been discussed in relation to suicidal behavior and alcoholism. Thus, it is of particular interest that alcoholism is a disorder associated with a very high risk of suicidal behavior. Kessel and Grossman (1961) estimated that the risk of completed suicide among male alcoholic individuals was 75- to 85-fold higher than that found in the general population. Kendell and Staton (1966) arrived at a figure of 58-fold, and Gillis (1969) estimated 60- to 70-fold. Miles (1977) reviewed 15 follow-up studies that reported the incidence of suicide among alcoholic individuals. He estimated that up to 15% of alcoholic individuals die by committing suicide, although some believe that this estimate is too high.

Studies that have examined the clinical correlates of suicidal behavior among alcoholic patients have yielded results pertinent to the theme that 5-HT may have a role in violence, impulsivity, and alcohol abuse. Approximately 80% of alcoholic individuals who commit suicide are male, and approximately 40% have exhibited previous suicidal behavior (reviewed in Roy and Linnoila 1986). Virkkunen (1971) compared 19 alcoholic patients who committed suicide while in psychiatric treatment with 21 nonalcoholic patients who similarly committed suicide while in treatment. All the suicide victims had been rated during their last admission to one psychiatric hospital. Virkkunen found that more of the alcoholic, than nonalcoholic, suicide victims had been rated as exhibiting either "markedly outwardly directed aggressiveness" (32% vs. 537, respectively) or "violent behaviour toward others" (26% vs. 10%, respectively). More recently, Berglund (1984) reported that significantly more of his alcoholic subjects who at follow-up had committed suicide, when compared with those who had not, had been rated at their first admission as showing irritability, dysphoria, or aggressiveness (30% vs. 21%, respectively; $P < .05$); as showing lability of affect, explosiveness, or affective incontinence (31% vs. 11%, respectively; $P < .01$); or as being brittle and sensitive (16% vs. 8%, respectively; $P < .05$).

### Postmortem Studies in Alcoholic Individuals Who Committed Suicide

Thus there is some evidence suggesting that the male alcoholic individual who commits suicide has exhibited prior suicidal and aggressive behavior. However, to date, there have been few postmortem studies examining 5-HT metabolite levels and receptors in the brains of alcoholic suicide victims.

In one such study Gottfries et al. (1975) measured the activity of MAO, an enzyme involved in 5-HT metabolism. They compared eight alcoholic suicide victims with 20 nonalcoholic control subjects and found significantly decreased MAO activity among the alcoholic suicide victims in all 13 parts of the brain studied. However, Cochran et al. (1976) found no significant differences for levels of 5-HT in any of 33 brain areas when six alcoholic suicide victims were compared with six depressed suicide victims and six healthy control subjects.

## SUBTYPES OF ALCOHOLISM

To date, CSF and postmortem studies of the serotonergic system have been carried out only in heterogeneous groups of alcoholic patients. This point is of relevance because there is increasing evidence that there may be meaningful subgroups of alcoholic patients. Such subgroups may be distinguishable in terms of age of onset, personality, genetic background, and symptoms.

For example, alcoholism, like other psychiatric disorders, tends to run in families. In an early review of the evidence for genetic factors in alcoholic patients, Goodwin (1971) concluded that "without known exception, every family study of alcoholism, irrespective of country of origin, has shown much higher rates of alcoholism among the relatives of alcoholics than apparently occurs in the general population" (p. 545). Since that review, adoption studies, particularly in Sweden, have strengthened the evidence that there may be genetic factors in alcoholism. These adoption studies have also led to the suggestion that there may be at least two distinct subtypes of alcoholism with differing genetic contributions (reviewed by Cloninger

1987, in press). Type I alcoholism is thought to be characterized by late onset, anxious (passive-dependent) personality traits, and the rapid development of tolerance and dependence on the antianxiety effects of alcohol. The expression of the genetic vulnerability to Type I alcoholism requires adverse environmental circumstance. Type II alcoholism is characterized by early onset, high genetic penetrance regardless of the quality of environment, antisocial personality traits, and the persistent seeking of alcohol for its euphoriant effects. This leads to the early onset of an inability to abstain entirely from alcohol, as well as fighting and arrests when drunk (Table 8-1).

Thus, alcoholism is increasingly viewed not as a discrete entity, but as heterogeneous in its etiology. Type I and Type II suggest that there may be at least two differing processes predisposing individuals to alcohol abuse. These developing typologies strongly suggest that alcoholism accompanied by antisocial behavior should be classified separately from alcoholism unrelated to antisocial behavior (Goodwin 1979). It has been suggested above that this finding may have relevance to the possible role of 5-HT in alcoholism. More specifically, 5-HT may have a greater role in Type II than in Type I alcoholism (reviewed in Roy et al. 1987b).

**Table 8-1.**  Distinguishing characteristics of two types of alcoholism

| Characteristic features | Types of alcoholism | |
|---|---|---|
| | Type I | Type II |
| **Alcohol-related problems** | | |
| Usual age of onset (years) | After 25 | Before 25 |
| Spontaneous alcohol seeking (inability to abstain) | Infrequent | Frequent |
| Fighting and arrests when drinking | Infrequent | Frequent |
| Psychological dependence (loss of control) | Frequent | Infrequent |
| Guilt and fear about alcohol dependence | Frequent | Infrequent |
| **Personality traits** | | |
| Novelty seeking | Low | High |
| Harm avoidance | High | Low |
| Reward dependence | High | Low |

*Source.* Reprinted from Cloninger R, *Science* 236:410–416, 1987, with permission of the American Association for the Advancement of Science. Copyright 1987, AAAS.

## RELATIONSHIP OF CEREBROSPINAL FLUID 5-HYDROXYINDOLEACETIC ACID LEVELS AND PERSONALITY

Some investigators have reported significant negative correlations between CSF 5-HIAA levels and various measures of personality. These studies too may be relevant to Type I and Type II subtypes of alcoholism. Brown et al. (1979) reported on 26 military men with various personality disorders who were undergoing inpatient evaluation of suitability for further military service. Brown et al. examined information from nine categories relating to a history of aggressive behavior and showed a significant negative correlation between a history of aggressive behavior and CSF 5-HIAA levels ($r = -.78$, P < .01). In these authors' second study, in which 12 subjects with borderline personality disorder were examined, the negative correlation between a history of aggression and CSF 5-HIAA levels was replicated (Brown et al. 1982). There was also a strong negative correlation between scores on the psychopathic deviance scale of the Minnesota Multiphasic Personality Inventory (MMPI) and CSF 5-HIAA levels ($r = -.77$, P < .04). These observations have recently been extended to include a significant negative correlation with self-rated childhood problems in these subjects (Brown et al. 1986).

In Sweden, Schalling et al. (1983) and Schalling and Åsberg (1985) reported a significant negative correlation between CSF 5-HIAA levels and psychoticism scores (i.e., aggressiveness, nonconformity) on the Eysenck Personality Questionnaire (EPQ) (Eysenck and Eysenck 1976). In suicidal patients these authors also found a negative correlation between CSF 5-HIAA levels and EPQ neuroticism scores (i.e., emotional instability). On the Karolinska Scales of Personality (KSP) low CSF 5-HIAA levels were associated with high scores for monotony avoidance and impulsivity as well as with low scores for socialization (Åsberg et al. 1987). Furthermore, van Praag (1986) reported that depressed patients with low CSF 5-HIAA levels, as compared to depressed patients with normal CSF 5-HIAA levels, had had significantly more contacts with police and arguments with relatives, spouse, colleagues, and friends; showed more hostility at interview; and had an impaired employment history due to arguments. Also, among healthy volunteers we recently noted a significant negative correlation between CSF 5-HIAA levels and scores on the "urge to act out hostility" subscale of Fould's Hostility Questionnaire (HDHQ) (Roy et al. 1988a).

### Antisocial Personality Traits and Alcoholism

In order to further test the role of personality in the etiology of alcoholism, Cloninger (in press) and colleagues performed a prospective study of 431 Swedish school children. Detailed interviews with their teachers when the children were 10 to 11 years of age allowed the investigators to perform personality ratings of these subjects as early adolescents. Independent records about alcohol abuse in these subjects as young adults, up to the age of 27 years, were also available. The investigators' prediction that early onset of alcohol abuse would be primarily Type II alcoholism and associated with antisocial personality traits (low harm avoidance, high novelty seeking, and low reward dependence) was confirmed. Parenthetically, subjects rated as average on these personality traits as early adolescents had low risk of alcoholism as young adults.

In another Swedish study, von Knorring et al. (1987) examined personality traits among adult male alcoholic patients. Subjects who met DSM-III criteria for alcohol dependence with onset before 30 years of age and with at least three severe social complications, including alcohol-related job loss or arrest, drunken driving, criminality, or illegal drug abuse, were classified as Type II alcoholic patients. Subjects with the onset of alcohol dependence after 30 years of age, or with fewer social complications, were classified as Type I alcoholic patients. The KSP were administered. Interestingly, von Knorring and his colleagues found that the Type II alcoholic patients had significantly higher scores than the Type I alcoholic patients on the Verbal Aggression Scales and the Impulsive Sensation-Seeking Psychopathy Factor, as well as significantly lower scores on the Inhibition of Aggression Scale. This group of investigators has also reported that platelet MAO activity is low among Type II alcoholic patients.

Schuckit (1986) recently reported a study that examined the clinical correlates of self-harm among primary alcoholic patients. He gave a structured questionnaire to 913 primary alcoholic patients entering an alcoholism treatment program in San Diego. Compared to other primary alcoholic patients, those with histories of suicide attempts (17%) were more likely to have had early life difficulties with police, school, and parents, and reported more misuse of drugs (Table 8-2). Schuckit concluded that although these problems were not severe enough to qualify for the diagnosis of antisocial personality disorder, their presence implies that an increased impulsivity may have antedated alcoholism in these men.

Thus, there is also evidence from these studies of antisocial and impulsive personality traits among alcohol abusers supporting the notion derived from family-genetic studies that there is an antisocial subtype of alcoholic individual.

## ROLE OF SEROTONIN IN AGGRESSIVE AND IMPULSIVE BEHAVIOR

### Animal Studies

There are also data from animal studies on the role of 5-HT in aggressive and impulsive behavior. Valzelli (1969, 1971) and co-workers, studying isolation-induced fighting in male mice and mouse-

**Table 8-2.**  Comparison of clinical correlates of self-harm among primary alcoholic patients with or without a history of suicide attempt

| | Group 1: Patients who had attempted suicide ($n = 155$) | Group 2: Patients who had *not* attempted suicide ($n = 758$) | $\chi^2$ |
|---|---|---|---|
| **Alcohol-related problems** | | | |
| Missed work | 78% | 66% | 7.596** |
| Fired | 41 | 32 | 4.433* |
| Auto accident | 61 | 45 | 11.866*** |
| Public intoxication arrest | 61 | 42 | 18.853*** |
| Separated/ divorced | 58 | 44 | 9.770** |
| **Antisocial problems before 16 years of age** | | | |
| Played truant | 55 | 47 | 17.621*** |
| Suspended/ expelled | 33 | 18 | |
| Considered incorrigible | 19 | 7 | 21.7   *** |
| Ran away repeatedly | 28 | 18 | 7.986** |
| Reform school | 5 | 2 | |
| Arrests | 11 | 6 | 4.402* |
| Injured someone in fight | 29 | 18 | 8.867** |
| Used weapon | 13 | 6 | 8.094** |
| **Drug use history** | | | |
| Marijuana | 49 | 40 | 3.904* |
| Hallucinogens | 15 | 9 | 4.511* |
| Opiates | 6 | 2 | 4.700* |

*$P < .05$.   **$P < .01$.   ***$P < .001$.

*Source.* Reprinted, with permission, from Schuckit MJ, *Journal of Studies on Alcohol* 47:78–81, 1986. Copyright 1986.

killing behavior in rats, observed an association between the emergence of these behaviors and a low 5-HIAA/5-HT ratio in whole-brain homogenates obtained from the aggressive animals. A low ratio is thought to reflect a low rate of 5-HT turnover. Reduction of CNS 5-HT by pharmacological or dietary means also led to increased shock-induced fighting or mouse-killing behavior in rats (Kantak et al. 1981; Katz 1980).

Mouse-killing behavior produced by depletion of 5-HT by $p$-chlorophenylalanine can be reduced by the 5-HT reuptake inhibitor fluoxetine (Berzenyi et al. 1983). Olfactory bulbectomy, which in mice leads to increased intraspecies aggression, produces 5-HT accumulation in the CNS. This accumulation could be indicative of reduced 5-HT turnover; moreover, the time courses of the 5-HT accumulation and the expression of the aggressive behavior parallel each other (Garris et al. 1984).

Both isolation- and olfactory bulbectomy-induced aggression can be reduced by 5-HT receptor agonists. 5-HT$_{1A}$ receptor agonists such as 8-hydroxy-2-(di-$n$-propylamino)tetralin (8-OH-DPAT) are more potent than 5-HT$_{1B}$ receptor agonists in these animal models (Molina et al. 1986). 5-HT itself, and 5-HT receptor agonists, show anatomical specificity in their effects of reducing mouse killing in rats.

These rodent models of aggressive behavior do not permit the differentiation of aggressiveness from impulsivity. However, Soubrie (1986) has persuasively argued that 5-HT serves as an endogenous inhibitor of the expression of various behaviors. Thus, reduced 5-HT function might not lead to aggressive behaviors per se, but would serve a permissive role for the ready expression of aggressive impulses. According to this model 5-HT depletion leads to a primary increase in impulsivity, and the increment in aggressiveness is one of the more observable manifestations of the behavioral change.

### Human Studies

*Reactive Hypoglycemic Tendency.* Virkkunen has also investigated glucose metabolism among the previously discussed violent and impulsive offenders (reviewed in Roy and Linnoila 1988). In the antisocial personality disorder and intermittent explosive disorder groups, blood glucose levels fell significantly below the nadir levels of the control subjects. Furthermore, a long duration of hypoglycemia in the violent offenders correlated with a low verbal IQ, the presence of tattoos and slashing scars, and behavioral and sleeping problems. There were also significant correlations between a long duration of hypoglycemia and truancy, stealing, crimes against property, having multiple sentences, and having a father who was violent when under the influence of alcohol.

Among the violent offenders with antisocial personality, insulin secretion during the oral glucose tolerance test (GTT) was found to be greater in those patients who also had an early undersocialized aggressive conduct disorder. Day-night rhythm disturbance, sleeping difficulties, and increasing behavioral problems during the evaluation were also more characteristic of subjects who developed high insulin peak values.

Also, 11 of the 20 previously discussed arsonists studied by Virkkunen et al. became hypoglycemic during the GTT (reviewed in Roy and Linnoila 1988). Among 24 healthy male volunteers Benton and Kumari (1982) noted significant correlations between the degree of hypoglycemia during the GTT and high scores on two psychological tests of hostility and frustration tolerance that were administered during the GTT.

Interestingly, Yamamoto et al. (1984a, 1984b, 1985) have demonstrated that lesions of the suprachiasmatic nucleus lead to rats that may be vulnerable to mild hypoglycemia. These rats show hyperinsulinemic and hypoglucagonemic responses to glucose and deoxyglucose challenges, respectively. The suprachiasmatic nucleus projects to the ventromedial nucleus of the hypothalamus. This hypothalamic area, together with the lateral hypothalamus, which is also connected to the suprachiasmatic nucleus (Swanson and Cowan 1977), participates in the control of feeding and satiety, particularly in regard to carbohydrate intake.

The suprachiasmatic nucleus is thought to be the major endogenous circadian pacemaker in the CNS (Moore and Eichler 1972). It receives a serotonergic input from the brain-stem raphe nuclei (Palkovits et al. 1977). Thus, this nucleus provides an anatomic link between 5-HT functions, regulation of circadian rhythms such as the sleep-wake cycle, and regulation of glucose metabolism.

*Y Chromosome.* A French study reported that violent arsonists with the XYY syndrome showed reduced accumulation of CSF 5-HIAA after probenecid loading (Bioulac et al. 1978, 1980). This finding is thought to be indicative of reduced 5-HT turnover in the CNS. Furthermore, Linnoila et al. (1983) had one impulsive violent offender with the XYY syndrome in their sample. This offender had one of the lowest CSF 5-HIAA concentrations. Also, a recent Danish study following XYY and XXY males in the community found that the XYY males exhibited increased aggression toward their spouses (Shiavi et al. 1988). These findings, together with findings from animal studies reporting an association between Y chromosome features and intermale aggression, and the pattern of inheritance of

Type II alcoholic individuals, have led us to postulate that the Y chromosome may contain factors important for the regulation of 5-HT turnover in men.

## CONCLUSIONS

A number of reports suggest that a deficiency in the serotonergic system may be associated with violent and suicidal behavior and alcohol abuse. Furthermore, hypoglycemia during the oral GTT and insomnia are very common among individuals with poor impulse control and low CSF 5-HIAA concentrations. Recent animal research has implicated the suprachiasmatic nucleus, which has a dense serotonergic innervation, as playing an important role in controlling both circadian rhythms and glucose metabolism. The insomnia of individuals with poor impulse control may be an indicator of circadian rhythm disturbance. Alcohol is known to release 5-HT acutely and chronically and thus to reduce presynaptic 5-HT stores (Ballenger et al. 1979). Thus we speculate that a functional serotonergic deficit may be conducive to poor impulse control, circadian rhythm disturbances, and glucose metabolism disturbances, and that these disturbances may be conducive to violent outbursts, suicide attempts, and Type II alcohol abuse (Figure 8-3).

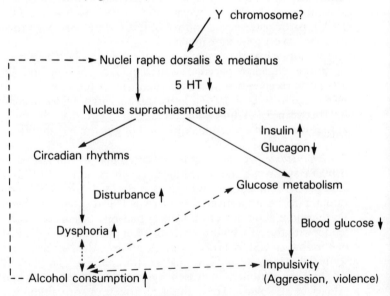

**Figure 8-3.** Schematic representation of proposed model for pathogenesis of impulsivity and Type II alcohol abuse.

Interestingly, preliminary research studies have reported that serotonergic medications may have some beneficial effects in alcoholic patients. Naranjo et al. (1985) found that among alcoholic patients, over the short term, the 5-HT reuptake blocker zimelidine increased the number of abstinent days, while on drinking days the amount of alcohol consumed was reduced. Further studies with various serotonergic drugs are in progress in different research centers. We believe that the time is also now ripe for double-blind, controlled studies of serotonergic drugs in the prevention of serious suicidal or violent behavior among individuals who have exhibited this behavior in the past.

# REFERENCES

Åsberg M, Träskman L, Thorén P: 5-HIAA in the cerebrospinal fluid: biochemical suicide predictor? Arch Gen Psychiatry 33:1193–1197, 1976

Åsberg M, Nordström P, Träskman-Bendz L: Biological factors in suicide, in Suicide. Edited by Roy A. Baltimore, MD, Williams & Wilkins, 1986, pp 47–71

Åsberg M, Schalling D, Träskman-Bendz L: Psychobiology of suicide, impulsivity and related phenomena, in Psychopharmacology: The Third Generation of Progress. Edited by Meltzer HY. New York, Raven Press, 1987, pp 665–668

Ballenger J, Goodwin F, Major L, et al: Alcohol and central serotonin metabolism in man. Arch Gen Psychiatry 36:224–227, 1979

Banki CM: Factors influencing monoamine metabolites and tryptophan in patients with alcohol dependence. J Neural Transm 50:98–101, 1981

Banki CM, Arato M: Amine metabolites and neuroendocrine responses related to depression and suicide. J Affective Disord 5:223–232, 1983

Benton D, Kumari N: Mild hypoglycemia and questionnaire measures of aggression. Biol Psychol 14:129–135, 1982

Berglund M: Suicide in alcoholism—a prospective study of 88 alcoholics: the multidimensional diagnosis at first admission. Arch Gen Psychiatry 41:888–891, 1984

Berzenyi P, Galateo E, Valzelli L: Fluoxetine activity on muricidal aggression induced in rats by *p*-chlorophenylalanine. Aggressive Behav 9:333–338, 1983

Bioulac B, Benezech M, Renaud B, et al: Biogenic amines in 47, XYY syndrome. Neuropsychopharmacology 4:366–370, 1978

Bioulac B, Benezech M, Renaud B, et al: Serotonergic dysfunction in the 47, XYY syndrome. Biol Psychiatry 15:917–923, 1980

Branchey L, Branchey M, Shaw S, et al: Depression, suicide and aggression in alcoholics and their relationship to plasma amino acids. Psychiatry Res 12:219–226, 1984

Brown G, Goodwin F, Ballenger J, et al: Aggression in humans correlates with cerebrospinal fluid metabolites. Psychiatry Res 1:131–139, 1979

Brown G, Ebert M, Goyer P, et al: Aggression, suicide, and serotonin: relationships to CSF amine metabolites. Am J Psychiatry 139:741–746, 1982

Brown G, Kline W, Goyer P, et al: Relationship of childhood characteristics to cerebrospinal fluid 5-hydroxy-indoleacetic acid in aggressive adults, in Proceedings of the IVth World Congress of Biological Psychiatry. Edited by Shagass C, et al. Amsterdam, Elsevier North-Holland, 1986, pp 177–179

Cloninger R: Neurogenetic adaptive mechanisms in alcohol. Science 236:410–416, 1987

Cloninger R: Genetic heterogeneity and the classification of alcoholism. Adv Alcohol Subst Abuse (in press)

Cochran E, Robins E, Grote S: Regional serotonin levels in brain, a comparison of depressive and alcoholic suicides with controls. Biol Psychiatry 11:283–294, 1976

Eichelman B: Role of biogenic amines in aggressive behaviors, in Psychopharmacology of Aggression. Edited by Sandler M. New York, Raven Press, 1979, pp 61–93

Eysenck HJ, Eysenck SB: Psychoticism as a Dimension of Personality. New York, Crane-Rusak, 1976

Garris D, Chamberlain J, DaVanzo J: Histofluorescent identification of indoleamine-concentrating brain loci associated with intraspecies, reflexive biting and locomotor behavior in olfactory-bulbectomized mice. Brain Res 294:385–389, 1984

Gillis L: The mortality rate and causes of death of treated chronic alcoholics. S Afr Med J 42:230–232, 1969

Goodwin D: Is alcoholism hereditary: a review and critique. Arch Gen Psychiatry 25:545–549, 1971

Goodwin D: Alcoholism and heredity: a review and hypothesis. Arch Gen Psychiatry 36:57–61, 1979

Gottfries C, Oreland L, Wiberg A, et al: Lowered monoamine oxidase activity in brains from alcohol suicides. Neurochemistry 25:667–673, 1975

Kantak K, Hegstrand L, Eichelman B: Facilitation of shock-induced fighting following intraventricular 5,7-dihydroxytryptamine and 6-hydroxy DOPA. Psychopharmacology (Berlin) 74:157–160, 1981

Katz R: Role of serotonergic mechanisms in animal models of predation. Prog Neuropsychopharmacol 4:219–231, 1980

Kendell R, Staton M: The fate of untreated alcoholics. Quarterly Journal of Studies on Alcohol 27:30–41, 1988

Kessel, N. Grossman G: Suicide in alcoholics. Br Med J 2:1671–1672, 1961

Lidberg L, Tuck J, Åsberg M, et el Homicide, suicide and CSF 5-HIAA. Acta Psychiatr Scand 71:230–236, 1985

Linnoila M, Virkkunen M, Scheinin M, et al: Low cerebrospinal fluid 5-hydroxyindoleacetic acid concentration differentiates impulsive from nonimpulsive violent behavior. Life Sci 33:2609–2614, 1983

Major L, Ballenger J, Goodwin F, et al: Cerebrospinal fluid homovanillic acid in male alcoholics: effect of disulfiram. Biol Psychiatry 12:635–642, 1987

Mann J, McBride P, Stanley M: Postmortem monoamine receptor and enzyme studies in suicide, in Psychobiology of Suicidal Behavior. Edited by Mann J, Stanley M. New York, New York Academy of Sciences, 1986a, pp 114–121

Mann J, Stanley M, McBride A, et al: Increased serotonin-2 and beta-adrenergic receptor binding in the frontal cortices of suicide victims. Arch Gen Psychiatry 43:954–959, 1986b

Miles C: Conditions predisposing to suicide: a review. J Nerv Ment Dis 164:231–248, 1977

Molina V, Gobaille S, Mandel P: Effects of serotonin-mimetic drugs on mouse-killing behavior. Aggressive Behav 12:201–211, 1986

Moore R, Eichler V: Loss of a circadian adrenal corticosterone rhythm following suprachiasmatic lesions in the rat. Brain Res 42:201–206, 1972

Naranjo C, Sellers E, Wu P: Moderation of ethanol drinking: role of enhanced serotonergic neurotransmission, in Research Advances in New Psychopharmacological Treatments for Alcoholism. Edited by Naranjo C, Sellers E. New York, Excerpta Medica, 1985, pp 171–186

Palkovits M, Saavedra J, Jacobovits D, et al: Serotonergic innervation of the forebrain: effect of lesions on serotonin and tryptophan hydroxylase levels. Brain Res 130:121–134, 1977

Rosenthal N, Davenport Y, Cowdry R, et al: Monoamine metabolites in cerebrospinal fluid of depressive subgroups. Psychiatry Res 2:113–119, 1980

Roy A, Linnoila M: Suicide and alcoholism, in Biology of Suicide. Edited by Maris R. New York, Guilford, 1986, pp 244–273

Roy A, Linnoila M: Suicidal behaviour impulsiveness and serotonin. Acta Psychiatr Scand 78:529–535, 1988

Roy A, Linnoila M: CSF studies on alcoholism and related behaviours. Progr Neuropsychopharmacol Biol Psychiatry 13:505–511, 1989

Roy A, Virkkunen M, Guthrie S, et al: Indices of serotonin and glucose metabolism in violent offenders, arsonists, and alcoholics, in Psychobiology of Suicidal Behavior. Edited by Mann J, Stanley M. Annals of the New York Academy of Sciences, Vol 487. New York, New York Academy of Sciences, 1986a, pp 202–220

Roy A, Virkkunen M, Guthrie S, et al: Monoamines, glucose metabolism, suicidal and aggressive behavior. Psychopharmacol Bull 22:661–665, 1986b

Roy A, Nutt D, Virkkunen M, et al: Serotonin, suicidal behavior and impulsivity. Lancet 2: 949–950, 1987a

Roy A, Virkkunen M, Linnoila M: Reduced central serotonin turnover in a subgroup of alcoholics. Prog Neuropsychopharmacol Biol Psychiatry 11:173–177, 1987b

Roy A, Adinoff B, Linnoila M: Acting out hostility in normal volunteers: negative correlation of 5-HIAA in cerebrospinal fluid levels. Psychiatry Res 24:187-194, 1988a

Roy A, Adinoff B, Roehrich L, et al: Pathological gambling: a psychobiological study. Arch Gen Psychiatry 45:369–373, 1988b

Roy A, Virkkunen M, Linnoila M: Monoamines, glucose metabolism, aggression towards self and others. Int J Neurosci 41:261–264, 1988c

Schalling D, Åsberg M: Biological and psychological correlates of impulsiveness and monotony avoidance, in The Biological Bases of Personality and Behavior, Vol 1. Edited by Strelau J, Farley F, Gale A. New York, Hemisphere, 1985

Schalling D, Edman G, Åsberg M: Impulsive cognitive style and inability to tolerate boredom: psychobiological studies of tempermental vulnerability, in Biological Bases of Sensation Seeking, Impulsivity and

Anxiety. Edited by Zuckerman M. Hillsdale, NJ, Erlbaum, 1983, pp 123–145

Schuckit M: Primary men alcoholics with histories of suicide attempts. J Stud Alcohol 47:78–81, 1986

Selmanoff M, Abreu E, Goldman D, et al: Manipulation of aggressive behavior in adult DBA/2/Bg and c57BL/10/Bg male mice implanted with testosterone in silastic tubing. Horm Behav 8:377–390, 1977

Shiavi R, Theilgaard A, Owen D, et al: Sex chromosome anomalies, hormones, and sexuality. Arch Gen Psychiatry 45:19–24, 1988

Stanley M, Mann J: Increased serotonin-2 binding sites in frontal cortex of suicide victims. Lancet 1:214–216, 1983

Stanley M, Mann J, Cohen S: Serotonin and serotonergic receptors in suicide, in Psychobiology of Suicidal Behavior. Edited by Mann J, Stanley M. Annals of the New York Academy of Sciences, Vol 487. New York, New York Academy of Sciences, 1986, pp 122–127

Soubrie P: Reconciling the role of central serotonin neurons in human and animal behavior. Behav Brain Sci 9:319–364, 1986

Swanson L, Cowan W: The efferent projections of the suprachiasmatic nucleus of the hypothalamus. J Comp Neurol 160:1–12, 1977

Takahashi S, Yamane H, Kondo H, et al: CSF monoamine metabolites in alcoholism, a comparative study with depression. Folia Psychiatrica Neurologica Japan 28:347–354, 1974

Träskman L, Åsberg M, Bertilsson L, et al: Monoamine metabolites in CSF and suicidal behavior. Arch Gen Psychiatry 38:631–636, 1981

Valzelli L: Aggressive behavior induced by isolation, in Excerpta Medica. Edited by Garattini S, Sigg E. Amsterdam, Excerpta Medica, 1969, pp 70–76

Valzelli L: Further aspects of the exploratory behavior in aggressive mice. Psychopharmacologia 19:91–94, 1971

van Praag H: Affective disorders and aggression disorders: evidence for a common biological mechanism. Suicide and Life-Threatening Behavior 16:21–50, 1986

Virkkunen M: Alcoholism and suicide in Helsinki. Psychiatr Fennica, 1971, pp 201–207

Virkkunen M, Nuutila A, Goodwin F, et al: CSF monoamine metabolites in male arsonists. Arch Gen Psychiatry 44:241–247, 1987

von Knorring I, von Knorring A, Smigan L, et al: Personality traits in subtypes of alcoholics. J Stud Alcohol 48:523–527, 1987

Wolkowitz O, Roy A, Doran A: Compulsive gambling and high risk sports, in Self Destructive Behavior. Philadelphia, PA, WB Saunders, 1985, pp 311–322

Yamamoto H, Nagai K, Nakagava H: Additional evidence that the suprachiasmatic nucleus is the center for regulation of insulin secretion and glucose homeostasis. Brain Res 304:237–241, 1984a

Yamamoto H, Nagai K, Nakagava H: Role of the suprachiasmatic nucleus in glucose homeostasis. Biomed Res 5:55–60, 1984b

Yamamoto H, Nagai K, Nakagava H: Lesions involving the suprachiasmatic nucleus eliminate the glucagon response to intracranial injection of 2-deoxy-D-glucose. Endocrinology 117: 468–473, 1985

Zuckerman M: Sensation seeking: a comparative approach to a human trait. Behav Brain Sci 7:413–471, 1984

Zuckerman M, Ballenger J, Jimerson D, et al: A correlational test in humans of the biological models of sensation seeking, impulsivity, and anxiety, in Biological Bases of Sensation Seeking, Impulsivity and Anxiety. Edited by Zuckerman M. Hillsdale, NJ, Erlbaum, 1983, pp 229–248

# Chapter 9

## *Serotonin in Schizophrenia*

John G. Csernansky, M.D.
Margaret Poscher, M.D.
Kym F. Faull, Ph.D.

# Chapter 9

## *Serotonin in Schizophrenia*

I n the 1950s D. W. Woolley first suggested that an abnormality of brain serotonin (5-hydroxytryptamine [5-HT]) function might play a role in the pathogenesis of schizophrenia (Woolley and Campbell 1962; Woolley and Shaw 1954). This theory was based upon the similarity in chemical structure between 5-HT and the hallucinogenic ergot alkaloids, and the observation that such compounds acted as antagonists of 5-HT in peripheral smooth muscle preparations. The psychotogenic effects of the most famous of these compounds, lysergic acid diethylamide (LSD), and of others, such as psilocybin, $N_1$, $N_4$-dimethyltryptamine (DMT), 5-methoxy-N, N-dimethyltryptamine, bufotenine, and the $\beta$-carbolines, led to two major lines of investigation.

First, it was proposed that these compounds might be synthesized in significant amounts in the brains of schizophrenic patients as the result of abnormal metabolic activity (i.e., methylation). For example, it was suspected that the hallucinogen DMT was being formed in the brains of schizophrenic patients in biologically active amounts. However, although DMT can be detected in human urine in minute amounts, no differences between schizophrenic patients and non-schizophrenic control subjects were found (see Rosengarten and Friedhoff 1976 for review). Also, there were important differences between the symptoms experienced after the administration of hallucinogenic compounds and those experienced during a schizophrenic psychosis (Hollister 1962, 1968). For example, although classic schizophrenic symptoms such as auditory hallucinations and paranoia follow intravenous administration of DMT,

This work was supported by a grant from the National Institute of Mental Health (MH-30854) to the Veterans Administration–Stanford Mental Health Clinical Research Center at Stanford University, and by a grant from the Research Service of the Veterans Administration (now Department of Veterans Affairs) to the Schizophrenia Biologic Research Center at the Palo Alto Veterans Administration Medical Center. The authors thank Pamela J. Elliott for manuscript preparation and editorial advice.

illusions and other characteristics of delirium are also common (Gillin et al. 1976a). Thus it became highly uncertain whether the psychosis induced by these compounds was a valid model for schizophrenia.

Second, the actions of these compounds on brain serotonergic systems led to hypotheses that an endogenous over- or underproduction of 5-HT might be related to the production of schizophrenic symptoms. The actions of LSD on brain 5-HT have been most widely studied. In classic work by several investigators, it was shown that LSD suppressed the firing rate of serotonergic neurons in the raphe nucleus, causing increases in brain 5-HT concentrations and decreases in 5-hydroxyindoleacetic acid (5-HIAA) concentrations (Aghajanian et al. 1968, 1970; Freedman 1961; Freedman and Giarman 1962). However, these changes cannot entirely account for the psychotogenic properties of LSD because of several important discrepancies. For example, the behavioral effects of LSD outlast its effect on the firing rate of serotonergic neurons, and behavioral tolerance develops to LSD but not to its effects on serotonergic neurons (McCall 1982). Currently, it appears that the effects of LSD and related compounds on brain serotonergic systems are complex, involving both presynaptic and postsynaptic serotonergic mechanisms (Freedman 1986). In-depth reviews of the pharmacology of hallucinogenics are available elsewhere (Barker et al. 1981; Boarder 1977; Brawley and Duffield 1972).

In addition to the findings discussed above linking 5-HT to schizophrenia, other important lines of evidence have supported this hypothesis, including relationships between cerebrospinal fluid (CSF) 5-HIAA concentrations and clusters of schizophrenic symptoms, changes in brain 5-HT receptors in postmortem schizophrenic brain, changes in platelet 5-HT content in schizophrenic patients, and serotonergic effects of drugs that improve schizophrenic symptoms. These findings will be emphasized in this chapter.

## BASIC ASPECTS OF PRIMATE SEROTONERGIC SYSTEMS RELEVANT TO SCHIZOPHRENIA

There are unique features of brain serotonergic systems that should be considered as we attempt to unravel the role of 5-HT in schizophrenia. 5-HT, the principal indole biogenic amine, is distributed widely in the central nervous system (CNS) in concentrations between 1 and 10 nmol/g of tissue. This pattern of distribution differs distinctly from that of the other major biogenic amine neurotransmitters, dopamine and norepinephrine, which tend to occur in discrete regions in higher concentrations. 5-HT cell bodies are restricted primarily to the midline pons and medulla, an area referred to as the raphe. Cell clusters from within the raphe give rise to four predominant serotonergic fiber tracts: the dorsal and ventral descending pathways that terminate in the lower brain stem and spinal cord; and the dorsal and ventral ascending pathways that course through the medial forebrain bundle and distribute widely to many brain regions of particular interest in schizophrenia, including the diencephalon, caudate nucleus, limbic forebrain, and frontal, cingulate, parietal, occipital, and temporal neocortex (Felten and Sladek 1983).

For a significant proportion of 5-HT–containing fibers, 5-HT release occurs in close proximity to 5-HT recognition sites on adjacent nerve cells. However, the opportunity to confirm this synaptic arrangement of cells awaits techniques with adequate resolution for autoradiographic, immuno-, or histochemical localization of both 5-HT and 5-HT receptors. Interestingly, in a significant number of brain regions, the correlation between 5-HT concentrations (or 5-HT reuptake activity) and 5-HT recognition sites is poor (Pazos et al. 1987a, 1987b). This finding leads to the hypothesis that in some brain regions, 5-HT may be released in areas of low receptor density, and have a "neurohumoral-like" effect. In this way 5-HT could have pervasive effects on a relatively large volume of brain tissue, effecting a constellation of processes that are not ordinarily thought of as being closely related in a neuroanatomical sense. One problem with such hypotheses concerns neurotransmitter dilution—at remote recognition sites would there be adequate concentrations to produce a response?

Regulation of synaptic 5-HT concentrations is achieved in much the same way as for other biogenic amine neurotransmitters. Active reuptake mechanisms exist and control of release is achieved at least in some fibers via autoreceptors on presynaptic terminals. 5-HT is metabolized via monoamine oxidase (MAO) and aldehyde dehydrogenase to form 5-HIAA. A small proportion (less that 5%) of the metabolized amine is excreted as 5-hydroxytryptophol. The rate of production of 5-HIAA is a reasonable index of the rate of turnover of 5-HT in the CNS (Neff and Tozer 1968; Neff et al. 1967). However, in clinical experiments, the rate of CNS 5-HIAA production can only be measured indirectly by assay of 5-HIAA in lumbar CSF, and the neuroanatomical origins of lumbar CSF 5-HIAA remain unclear. A significant proportion must come from metabolism in the spinal cord, as well as brain (for review, see Scheinin 1985). Direct measurement of 5-HT concentrations in CSF is difficult because the parent amine is present in about 0.02% of 5-HIAA concentrations and is below the limit of detection of most assays. Measurements of 5-HIAA concentrations in plasma and urine largely reflect 5-HT metabolism in peripheral tissues, particularly the gut, where 80% of total body 5-HT occurs in enterochromaffin cells.

Finally, it is naive to assume that a disease as complex as schizophrenia with multiple symptoms and signs could be caused by a defect in a single neurotransmitter system. Interrelationships among different CNS neurotransmitter systems almost certainly impact on the course of the disease and the manifested symptoms. Defining the functional and anatomical interrelationships between 5-HT and other neurotransmitter systems is an important area of current investigation. Complex interactions between dopamine and 5-HT have already been demonstrated (Balsara et al. 1979; Blackburn et al. 1982; Carter and Pycock 1978; Korsgaard et al. 1985; Lee and Geyer 1982). Colocalization of 5-HT in the same neurons with peptide neurotransmitters, including substance P and thyrotropin-releasing hormone (TRH), has been demonstrated in the lower brain stem and spinal cord (Hokfelt et al. 1980). Given the expectation that more examples of interrelationships between 5-HT and other neurotransmitters will soon be described, the tendency to concentrate on single systems will require revision.

## PSYCHOBIOLOGIC STUDIES OF SEROTONIN IN SCHIZOPHRENIA

### Peripheral Serotonin Markers

Because of the problems inherent with obtaining direct measures of CNS 5-HT activity in clinical populations, a number of investigators have resorted to using peripheral markers of 5-HT activity. Human blood platelets are the most widely adopted peripheral model of CNS amine-containing neurons. Several biochemical analogies between blood platelets and CNS neurons provide the basis for studying platelet biochemical processes in patients with schizophrenia (Stahl et al. 1982). A recurring finding has been elevated platelet 5-HT concentrations in schizophrenic patients (for review, see Meltzer 1987). However, elevated platelet 5-HT concentrations are not specific for schizophrenia; they are also found in mood disorders, organic brain disease (especially mental retardation), and childhood autism (Partington et al. 1973). The biochemical basis for elevated platelet 5-HT concentrations in these disorders remains unclear. Furthermore, different mechanisms may apply in each case (Stahl et al. 1983). However, irrespective of this nonspecificity, these findings are of interest because they do provide a link between seemingly different forms of mental disorder.

## Cerebrospinal Fluid 5-HIAA Concentrations and Symptom Correlations

Despite the uncertainty regarding the origin of 5-HIAA in lumbar CSF, many investigators have chosen to study lumbar CSF 5-HIAA concentrations as an index of CNS 5-HT function. CSF 5-HIAA concentrations at baseline and following probenecid administration have been examined and have yielded similar findings. CSF 5-HIAA concentrations have not shown consistent differences between schizophrenic and comparison groups (Berger et al. 1980; Post et al. 1975). However, within groups of schizophrenic patients, significant differences have been found. As early as 1966, Ashcroft et al. found that CSF 5-HIAA concentrations were lower in "acute" compared to "chronic" schizophrenic patients. However, it is difficult to make the criteria for such course of illness designations explicit, and state versus trait issues remain unresolved. Other investigators have also found CSF 5-HIAA concentrations to be significantly lower in schizo-phrenic patients when they are in the acute stages of their illness (Bowers 1978; Bowers et al. 1969; Subrahmanyam 1975). More recently, many investigators have begun to refine their hypotheses and search for relationships between CSF 5-HIAA concentrations and specific aspects of schizophrenic symptomatology. Current indica-tions are that this approach is meeting with some success and could provide a basis for continued investigation.

The most widely cited achievement has been by Åsberg in Sweden, and now confirmed by others, who found evidence of a correlation between low CSF 5-HIAA and an increased risk for violent suicide (Åsberg et al. 1976). Although initially found in depressed patients, this correlation has now been independently demonstrated in schizophrenic patients with violent attempts or completed suicides (Banki et al. 1984; Ninan et al. 1984). Interestingly, several inves-tigators have also found a relationship between lower CSF 5-HIAA concentrations and positive symptoms or agitation, phenomena that are known to predispose to acts of violence in schizophrenic patients (Yesavage 1984). For example, although Post et al. (1975) did not detect any group difference in CSF 5-HIAA concentrations between schizophrenic patients and control subjects, they did find an inverse correlation between CSF 5-HIAA concentrations and the severity of hallucinations and a measure of Schneiderian first-rank symptoms. Similarly, Lindstrom (1985) found an inverse correlation between CSF 5-HIAA concentrations and the severity of delusions and sadness in schizophrenic patients. Bowers (1978) demonstrated inverse cor-relations between CSF 5-HIAA concentrations after probenecid ad-ministration and subjective assessments of psychomotor agitation and numerical counts of movement in a group of 10 schizophrenic patients (Figure 9-1). In this same study, Bowers found an inverse correlation between CSF 5-HIAA concentrations and good prognosis. In summary, the CSF 5-HIAA concentration data suggest that reduced CNS 5-HT function may be linked to positive symptoms, agitation, self-harm, and a more acute course of illness in schizophrenia.

Other investigations have suggested that CSF 5-HIAA concentrations are related to measures of negative schizophrenic symptoms and the deficit syndrome. In addition to the inverse relationships found between CSF 5-HIAA concentrations and measures of psychomotor agitation, Bowers (1978) also reported a significant direct correlation

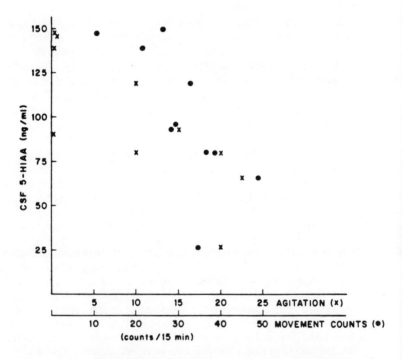

**Figure 9-1.**  CSF 5-HIAA, movement counts, and motor agitation rating in newly admitted schizophrenic patients. Movement counts, $r = -.72$ ($P < .05$); agitation, $r = -.79$ ($P < .01$). Reproduced from Bowers MB Jr, in *Biochemistry of Mental Disorders*. Edited by Usdin E, Mandell A. New York, Marcel Dekker, 1978, p 199, with permission of Marcel Dekker. Copyright 1978, Marcel Dekker.

between emotional withdrawal assessed with the Brief Psychiatric Rating Scale (BPRS) and CSF 5-HIAA concentrations following probenecid administration. More recently, our group has shown that two measures of 5-HT function, CSF 5-HIAA concentrations and platelet 5-HT concentrations, correlate directly with autistic mannerisms and posturing in schizophrenic patients (Figure 9-2) (King et al. 1985). At the Schizophrenia Biologic Research Center, mannerisms and posturing covary with the other more classical manifes-

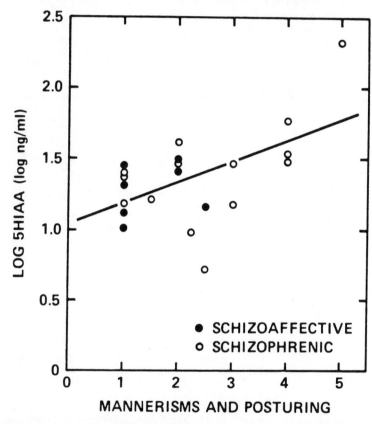

**Figure 9-2.** Scatterplot of log CSF 5-HIAA concentrations vs. mannerisms and posturing as assessed by the Brief Psychiatric Rating Scale for schizophrenic and schizoaffective subjects. $r = .56$; $P < .01$. Reproduced from King et al., *Psychiatry Research* 14:235–240, 1985, with permission of Elsevier Science Publishing Company. Copyright 1985, Elsevier Science Publishing Company.

tations of the deficit syndrome (Guelfi et al. 1989). In fact, we have now extended the finding of King et al. (1985) with the observation that CSF 5-HIAA concentrations correlate directly with several BPRS items that reflect negative symptoms, including mannerisms and posturing, poor work history as measured with the Strauss-Carpenter Scale (Strauss and Carpenter 1974), and impairment on three Wechsler Adult Intelligence Scale—Revised (WAIS-R) subscale scores that test attention. To avoid the estimation of many correlation coefficients, we combined these measures into a single factor score using principal components analysis (Figure 9-3) (Csernansky et al., in press). In summary, these findings add support to the hypothesis that brain 5-HT function is linked to deficit schizophrenic characteristics. However, they also raise the question as to whether agitation

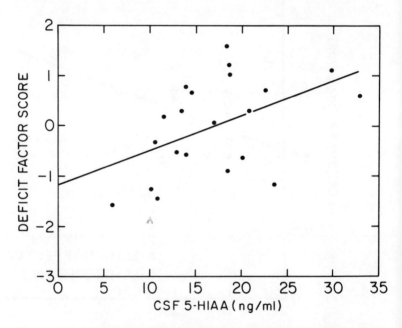

**Figure 9-3.** Scatterplot of CSF 5-HIAA concentrations vs. deficit factor scores in 21 drug-free schizophrenic patients. Each factor score combined data from four Brief Psychiatric Rating Scale items (emotional withdrawal, mannerisms and posturing, blunted affect, and motor retardation), the work history item of the Strauss-Carpenter Scale, and three Wechsler Adult Intelligence Scale—Revised subscales (digit symbol, digit span, and picture arrangement). $r = .50$; $N = 21$; $P < .05$.

and positive symptoms in schizophrenic patients occur on a continuum with the deficit syndrome, with CNS 5-HT function being a neurochemical correlate along the entire length. In general, positive and negative symptoms do not occur reciprocally in schizophrenic patients (for review, see Guelfi et al. 1989). Yet, the techniques required to assess behavioral syndromes linked to under- and over-production of CNS 5-HT should go beyond the measurement of symptom patterns at one point in time, and include variables related to cognition, psychosocial function, and course of illness.

### Postmortem Findings

No significant differences between schizophrenic patients and control groups have been found with respect to concentrations of 5-HT, 5-HIAA, and tryptophan in postmortem CNS tissue (Joseph et al. 1979). However, studies on the number and affinity of 5-HT recognition sites in autopsied tissue are an area of promising research. Several years ago, Bennett et al. (1979) demonstrated 40–50% decreases in the binding of $^3$H-labeled LSD (a ligand that identifies several 5-HT receptor subtypes) in autopsied brain tissue from schizophrenic individuals. Mann et al. (1986) found a significant 28% increase in the density of 5-HT$_2$ receptors in the frontal cortex of suicide victims, although the diagnoses of the individuals were not specified. Recently, there has been a proliferation of specific ligands for different subtypes of 5-HT$_1$, 5-HT$_2$, and 5-HT$_3$ receptors (Fozard 1987). The availability of these new pharmacological probes should be exploited to extend these findings. If the CSF 5-HIAA data in schizophrenia reflect long-standing alterations in CNS 5-HT turn-over, reciprocal changes in 5-HT receptors might accompany these alterations. In parallel with CSF 5-HIAA studies, relationships should be sought between alterations in CNS 5-HT receptor densities and specific dimensions of psychopathology.

## PSYCHOPHARMACOLOGIC TRIALS OF SEROTONERGIC AGENTS IN SCHIZOPHRENIA

Therapeutic trials of serotonergic drugs should be evaluated in light of the fact that much remains to be learned about the fundamentals of CNS serotonergic pharmacology. The functions of all 5-HT receptor subtypes have yet to be fully characterized. In particular, cellular locations (e.g., postsynaptic vs. presynaptic) and neurophysiological actions remain under intense investigation (for review, see Bradley et al. 1986). In addition, since 5-HT interacts with other neurotransmitters that may have important roles in the pathogenesis of schizophrenia, the therapeutic actions of serotonergic compounds could occur via indirect effects. For example, in the striatum, presynaptic 5-HT heteroreceptors modulate the release of dopamine (for review, see Chesselet 1984).

## Serotonin Precursor and Depleting Agents in Schizophrenia

The treatment of schizophrenic symptoms has been attempted by both enhancing and inhibiting 5-HT transmission. The simultaneous pursuit of these two strategies reflects the fact that during the original investigation of hallucinogenic 5-HT compounds, these drugs were purported to act as both 5-HT agonists and antagonists. Pursuing the first strategy of enhancing CNS 5-HT function, Klee et al. (1960) administered large doses of the 5-HT precursor 5-hydroxytryptophan (5-HTP) to schizophrenic patients. However, no patients improved, while prominent dose-dependent side effects (i.e., nausea) occurred as the result of peripheral metabolism of 5-HTP. More recently, Bigelow et al. (1979) made several attempts to infuse 5-HTP together with a 1-aromatic acid decarboxylase inhibitor. This combination was selected in order to lessen the amount of 5-HTP available for peripheral metabolism. However, these investigators still found no improvement. In fact, one patient treated with both 5-HTP and chlorpromazine worsened. Yet, as with many attempts to test novel drug treatments, the patients chosen for study were known to be refractory to all other treatments. Gillin et al. (1976b) treated eight patients with orally administered L-tryptophan and found that in three patients the symptoms increased, in only one they decreased, and in four no change was found. To date, the strategy of enhancing CNS 5-HT function has not been tested in schizophrenic patients known to be responsive to conventional neuroleptic therapy.

Attempts to treat schizophrenic patients by decreasing 5-HT function have also been inconclusive. p-Chlorophenylalanine, which inhibits tryptophan hydroxylase, can be used to globally decrease CNS 5-HT function (for review, see Sjoerdsma et al. 1970). DeLisi et al. (1982) reported on the effects of p-chlorophenylalanine in seven neuroleptic nonresponsive schizophrenic patients, selected for having been made worse by 5-HTP administration or for having high whole-blood 5-HT concentrations. Although no substantial improvement in psychosis was seen, an increase in socialization was observed in some patients. Similarly, with the recent increasing emphasis on negative schizophrenic symptoms, Stahl et al. (1985) attempted to specifically treat negative symptoms in 12 schizophrenic patients with fenfluramine, an amphetamine analogue that depletes 5-HT after chronic administration. Three patients improved, although no effect on the overall group could be demonstrated. Shore et al. (1985) could demonstrate no improvement with fenfluramine in eight refractory schizophrenic patients; however, a specific effect on negative symptoms was not sought.

## Serotonin Receptor Antagonists in Schizophrenia

More recently, attempts have been made to treat the negative symptoms of schizophrenia with neuroleptics ($D_2$ antagonists) that are also direct $5\text{-}HT_2$ antagonists, such as setoperone (Ceulemans et al. 1985). Forty patients, selected for having prominent negative symptoms, were studied for this preliminary open trial, and a 50% improvement in various types of symptoms, including emotional withdrawal, hallucinations, and depression, was noted. However, since $5\text{-}HT_2$ receptor antagonists might also have antidepressant effects, these results should be interpreted with caution. Negative schizophrenic symptoms remain difficult to distinguish from the vegetative features of depression (Prosser et al. 1987). Other direct $5\text{-}HT_2$ antagonists, such as ritanserin, have also been tested for their effects on negative symptoms in schizophrenic patients, with promising results (for review, see Bleich et al. 1988). However, in addition to being $5\text{-}HT_2$ antagonists, this class of drugs, as exemplified by ketanserin, may also act within biogenic amine terminals to alter the release of neurotransmitters and their metabolites (Leysen et al. 1988). Thus, there remains a need for completely specific $5\text{-}HT_2$ antagonists to be used as antischizophrenic drugs in testing $5\text{-}HT$ hypotheses.

Many other conventional and atypical neuroleptics (e.g., the butyrophenones and clozapine) are potent $5\text{-}HT_2$ antagonists. However, in a comparison of clinical potency versus binding affinity at several neurotransmitter receptors, no correlation was seen between clinical potency and binding affinity to $5\text{-}HT_2$ receptors (Peroutka and Snyder 1980). Interestingly, in this study, clinical potency was defined with the usual emphasis on positive schizophrenic symptoms. The relative potency of neuroleptics for the treatment of negative symptoms is not generally known, and a stronger rank-order correlation between clinical potency and $5\text{-}HT_2$ binding affinity might well be obtained if the response of negative symptoms is emphasized. Determining the relationship between clinical potency to treat negative symptoms and binding affinity to $5\text{-}HT_2$ receptors could be revealing as new drugs that act primarily as $5\text{-}HT_2$ antagonists are tested in clinical studies.

Lastly, a new approach in treating schizophrenia may result from the recent characterization of a new $5\text{-}HT$ receptor subtype in brain, termed $5\text{-}HT;i3$. Interestingly, the $5\text{-}HT;i3$ receptor subtype corresponds to the serotonin-M receptor, discovered in peripheral mammalian tissues more than 30 years ago (Richardson and Engel 1986). Some familiar compounds, such as metaclopromide and (-)-cocaine, are ligands for the 5-HT3 receptor, and specific 5-HT3 agents have only recently been synthesized. Studies of 5-HT3 receptor binding, using specific ligands, have been conducted in peripheral mammalian tissues (Bradley et al. 1986) and rat brain (Kilpatrick et al. 1987). Studies in postmortem human brain are underway (e.g., Costall and Naylor 1988; Tyers 1988). Unlike other 5-HT receptor subtypes, the 5-HT3 receptor is a component of a cation channel (Derkach et al. 1989). 5-HT3 receptors may play an important role in regulating the release of other neurotransmitters (Barnes et al. 1989).

One of the new selective 5-HT3 antagonists, GR38032F (Brittain et al. 1987), has already been shown to have antipsychotic-like activity in an animal model of amphetamine-induced hyperlocomotion (Costall et al. 1987a). Preliminary clinical studies may soon follow once the toxicology of these compounds in humans has been more fully investigated. However, given these preliminary animal data, and the fact that hyperlocomotion in rodents has often been linked to mesolimbic dopamine function, an interaction of 5-HT3 receptors with mesolimbic dopamine mechanisms should be expected (Costall et al. 1987b). Perhaps patients with a history of good clinical responses to antidopaminergic neuroleptics should be chosen as the first subjects for trials with 5-HT3 antagonists.

## FUTURE DIRECTIONS

Studies of 5-HT function in schizophrenia should not be investigated or interpreted in a vacuum. Recent studies have suggested that CSF 5-HIAA concentrations may be correlated with both structural changes and genetic factors in schizophrenia. Three groups of investigators have reported that CSF 5-HIAA concentrations are inversely correlated with the ventricular-brain ratio (VBR) (Nyback et al. 1984; Potkin et al. 1983; van Kammen et al. 1985). In addition, van Kammen et al. (1985) reported that CSF 5-HIAA concentrations correlated inversely with evidence of cortical atrophy. Hopefully, these relationships are reflecting the loss of particular brain areas innervated by 5-HT projections. However, larger total CSF volumes could nonspecifically result in lower CSF 5-HIAA concentrations, with no absolute change in the absolute amount of 5-HIAA produced (Scheinin 1985). Structural changes have been found in a variety of brain areas in schizophrenia, including the cortex and the hippocampus (for review, see Kovelman and Scheibel 1986). These two areas are highly innervated by 5-HT projections. In future structural studies of autopsied brain tissue from schizophrenic individuals, efforts should be made to find relationships between specific changes in such brain areas and measures of 5-HT function.

Sedvall and Wode-Helgodt (1980) and Lindstrom (1985) have reported that CSF 5-HIAA concentrations are higher in schizophrenic patients with a family history of the disorder, and in nonschizophrenic individuals with a family history of schizophrenia (Sedvall et al. 1980). This finding suggests that measurements of 5-HT function should be included among the biological variables chosen for study in schizophrenic pedigrees. Furthermore, relationships should be sought between CSF 5-HIAA concentrations and other biological phenomena (e.g., eye-tracking deficits) that are thought to identify inherited vulnerability factors in schizophrenia.

Future studies of 5-HT in schizophrenia should also include other functional 5-HT measures. So far, only two measures of 5-HT metabolism, CSF 5-HIAA and platelet 5-HT, have been extensively investigated in schizophrenic patients. While neuroendocrine studies of 5-HT function have yielded significant advances in our understanding of serotonergic mechanisms in both depression and anxiety disorders, few neuroendocrine studies have been performed in schizophrenia to specifically test serotonergic hypotheses. Abnormalities of neuroendocrine function that might reflect the role of 5-HT, such as dexamethasone nonsuppression (Banki et al. 1984) or prolactin secretion (for review, see Meltzer 1984), have been studied in schizophrenic patients. Similar studies with a more specific focus on serotonergic mechanisms might provide an interesting counterpoint to CSF 5-HIAA and platelet 5-HT measurements. Also, highly specific serotonergic ligands labeled with positron-emitting radionuclides could be used to study $5-HT_2$ receptor densities in selected brain areas in schizophrenic patients. Such studies could confirm and extend postmortem receptor studies of autopsied brain tissue from schizophrenic individuals. Finally, relationships among $5-HT_2$ receptor density, CSF 5-HIAA concentrations, and neuroendocrine 5-HT measures would be of interest.

## CONCLUSIONS

Studies of 5-HT continue to contribute to our understanding of possible pathogenetic mechanisms in schizophrenia. Relative decreases in CNS 5-HT function may be linked to positive symptoms and increased acuity in schizophrenic patients, while relative increases are linked to deficit characteristics. Postmortem studies of receptor neurochemistry as well as novel drug studies can now be pursued using more specific pharmacological probes of the various 5-HT receptor subtypes. However, it seems unlikely that schizophrenia will be understood as a single lesion affecting only one neurotransmitter system. It is important to continue to develop pathogenetic models of this disease that emphasize the interrelationships among several neurotransmitter systems, including 5-HT. The role of 5-HT in the functioning of specific brain areas of high interest in schizophrenia, such as the cortex, the hippocampus, and the basal ganglia, should also be emphasized.

## *REFERENCES*

Aghajanian GK, Foote WE, Sheard MH: Lysergic acid diethylamide: sensitive neuronal units in the midbrain raphe. Science 161:706–708, 1968

Aghajanian GK, Foote WE, Sheard MH: Action of psychotogenic drugs on midbrain raphe neurons. J Pharmacol Exp Ther 171:178–187, 1970

Åsberg M, Thorén P, Träskman L: 5-HIAA in the cerebrospinal fluid—a biochemical suicide predictor? Arch Gen Psychiatry 33:1193–1197, 1976

Ashcroft GW, Crawford TBB, Eccleston D, et al: 5-Hydroxyindole compounds in the cerebrospinal fluid of patients with psychiatric or neurological diseases. Lancet 2:1049–1052, 1966

Balsara JJ, Jadhav JH, Chandorkar AG: Effect of drugs influencing central serotonergic mechanisms on haloperidol-induced catalepsy. Psychopharmacology (Berlin) 62:67–69, 1979

Banki CM, Arato M, Papp Z, et al: Cerebrospinal fluid amine metabolites and neuroendocrine changes in psychoses and suicide, in Catecholamines: Neuropharmacology and Central Nervous System— Therapeutic Aspects. Edited by Usdin E, Carlsson A, Dahlstrom A. New York, Alan R Liss, 1984, pp 153–159

Barker SA, Monti JA, Christian ST: N,N-Dimethyltryptamine: an endogenous hallucinogen. Int Rev Neurobiol 22:83–110, 1981

Barnes JM, Barnes NM, Costall B, et al: 5-HT3 receptors mediate inhibition of acetylcholine release in cortical tissue. Nature 338:762–763, 1989

Bennett JP Jr, Enna SJ, Bylund DB, et al: Neurotransmitter receptors in frontal cortex of schizophrenics. Arch Gen Psychiatry 36:927–934, 1979

Berger PA, Faull KF, Kilkowski J, et al: CSF monoamine metabolites in depression and schizophrenia. Am J Psychiatry 137:174–180, 1980

Bigelow LB, Walls P, Gillin JC, et al: Clinical effects of L-5-hydroxytryptophan administration in chronic schizophrenic patients. Biol Psychiatry 14:53–67, 1979

Blackburn TP, Cox B, Lee TF: Involvement of a central dopaminergic system in 5-methoxy-N, N-dimethyltryptamine-induced turning behavior in rats with lesions of the dorsal raphe nuclei. Psychopharmacology (Berlin) 78:261–265, 1982

Bleich A, Brown SL, Kahn R, et al: The role of serotonin in schizophrenia. Schizophr Bull 14:297–315, 1988

Boarder MR: The mode of action of indoleamine and other hallucinogens, in Essays in Neurochemistry and Neuropharmacology, Vol 2. Edited by Youdim MBH, Lovenberg W, Sharman DF, et al. New York, John Wiley, 1977, pp 21–48

Bowers MB Jr: Serotonin systems in psychotic states, in Biochemistry of Mental Disorders. Edited by Usdin E, Mandell A. New York, Marcel Dekker, 1978, pp 191–204

Bowers MB Jr, Heninger GR, Gerbode F: Cerebrospinal fluid 5-HIAA and HVA in psychiatric patients. Int J Neuropharmacol 8:255–262, 1969

Bradley PB, Engel G, Feniuk W, et al: Proposals for the classification and nomenclature of functional receptors for 5-hydroxytryptamine. Neuropharmacology 25:563–576, 1986

Brawley P, Duffield JC: The pharmacology of hallucinogens. Pharmacol Rev 24:31–66, 1972

Brittain RT, Butler A, Coates IH, et al: GR38032F, a novel selective 5HT$_3$ receptor antagonist. Br J Pharmacol 90:87P, 1987

Carter CJ, Pycock CJ: Differential effects of central serotonin manipulation on hyperactive and stereotyped behavior. Life Sci 23:953–960, 1978

Ceulemans DLS, Gelders YG, Hoppenbrouwers MLJA, et al: Effect of serotonin antagonism in schizophrenia: a pilot study with setoperone. Psychopharmacology (Berlin) 85:329–332, 1985

Chesselet MF: Presynaptic regulation of neurotransmitter release in the brain. Neuroscience 12:347–375, 1984

Costall B, Naylor R: The dopamine infusion model. Paper presented at the Fourth Biennial Winter Workshop on Schizophrenia, Badgastein, Austria, January 1988

Costall B, Domeney AM, Kelly ME, et al: The antipsychotic potential of GR38032F, a selective antagonist of 5HT$_3$ receptors in the central nervous system. Br J Pharmacol 90:89P, 1987a

Costall B, Domeney AM, Naylor RJ, et al: Effects of the 5-HT$_3$ receptor antagonist, GR38032F, on raised dopaminergic activity in the mesolimbic system of the rat and marmoset brain. Br J Pharmacol 92:881–894, 1987b

Csernansky JG, King RJ, Faustman WO, et al: CSF 5-HIAA and deficit schizophrenic characteristics. Br J Psychiatry (in press)

DeLisi LE, Freed WJ, Gillin JC, et al: *p*-Chlorophenylalanine trials in schizophrenic patients. Biol Psychiatry 17:471–477, 1982

Derkach V, Surprenant A, North RA: 5-HT3 receptors are membrane ion channels. Nature 339:706–709, 1989

Felten DL, Sladek JR Jr: Monoamine distribution in primate brain. V. Monoaminergic nuclei: anatomy, pathways and local organization. Brain Res Bull 10:171–284, 1983

Fozard JR: 5-HT: the enigma variations. Trends in Pharmacological Sciences 8:51–56, 1987

Freedman DX: Effects of LSD-25 on brain serotonin. J Pharmacol Exp Ther 134:160–166, 1961

Freedman DX: Hallucinogenic drug research—if so, so what?: symposium summary and commentary. Pharmacol Biochem Behav 24:407–415, 1986

Freedman DX, Giarman NJ: LSD-25 and the status and level of brain serotonin. Ann NY Acad Sci 90:98–107, 1962

Gillin JC, Kaplan J, Stillman R, et al: The psychedelic model of schizophrenia: the case of N,N-dimethyltryptamine. Am J Psychiatry 133:203–208, 1976a

Gillin JC, Kaplan JA, Wyatt RJ: Clinical effects of tryptophan in chronic schizophrenic patients. Biol Psychiatry 11:635–639, 1976b

Guelfi GP, Faustman WO, Csernansky JG: Independence of positive and negative symptoms in a population of schizophrenic patients. J Nerv Ment Dis 177(5):285–290, 1989

Hokfelt T, Lundberg JM, Schultzberg M, et al: Co-existence of peptides and putative transmitters in neurons. Adv Biochem Psychopharmacol 22:1–23, 1980

Hollister LE: Drug-induced psychoses and schizophrenic reactions: a critical comparison. Ann NY Acad Sci 96:80–92, 1962

Hollister LE: Chemical Psychoses: LSD and Related Drugs. Springfield, IL, Charles C Thomas, 1968

Joseph MH, Baker HF, Crow TH, et al: Brain tryptophan metabolism in schizophrenia: a post mortem study of metabolites on the serotonin and kynurenine pathways in schizophrenic and control subjects. Psychopharmacology (Berlin) 62:279–285, 1979

King R, Faull KF, Stahl SM, et al: Serotonin and schizophrenia: correlations between serotonergic activity and schizophrenic motor behavior. Psychiatry Res 14:235–240, 1985

Kilpatrick GJ, Jones BJ, Tyers MB: Identification and distribution of 5-HT3 receptors in rat brain using radioligand binding. Nature 330:746–748, 1987

Klee GD, Bertino J, Goodman A, et al: The effects of 5-hydroxytryptophan (a serotonin precursor) in schizophrenic patients. J Ment Sci 106:309–316, 1960

Korsgaard S, Gerlach J, Christensson E: Behavioral aspects of serotonin-dopamine interaction in the monkey. Eur J Pharmacol 118:245–252, 1985

Kovelman JA, Scheibel AB: Biological substrates of schizophrenia. Acta Neurol Scand 73:1–32, 1986

Lee EHY, Geyer MA: Selective effects of apomorphine on dorsal raphe neurons: a cytofluorimetric study. Brain Res Bull 9:719–725, 1982

Leysen JE, Eeens A, Gommeren W, et al: Identification of nonserotonergic [$^3$H]ketanserin binding sites associated with nerve terminals in rat brain and with platelets: relation with release of biogenic amine metabolites induced by ketanserin- and tetrabenazine-like drugs. J Pharmacol Exp Ther 244:310–321, 1988

Lindstrom LA: Low HVA and normal 5-HIAA CSF levels on drug-free schizophrenic patients compared to healthy volunteers: correlations to symptomatology and family history. Psychiatry Res 14:265–273, 1985

Mann JJ, Stanley M, McBride A, et al: Increased serotonin2 and β-adrenergic receptor binding in the frontal cortices of suicide victims. Arch Gen Psychiatry 43:954–959, 1986

McCall RB: Neurophysiological effects of hallucinogens on serotonergic neuronal systems. Neurosci Biobehav Rev 6:509–514, 1982

Meltzer HY: Neuroendocrine abnormalities in schizophrenia: prolactin, growth hormone, and gonadotropins, in Neuroendocrinology and Psychiatric Disorder. Edited by Brown GM. New York, Raven Press, 1984, pp 1–28

Meltzer HY: Biological studies in schizophrenia. Schizophr Bull 13:77–111, 1987

Neff NH, Tozer TN: In vivo measurement of brain serotonin turnover. Adv Pharmacol 6:97–109, 1968

Neff NH, Tozer TN, Brodie BB: Applications of steady-state kinetics to studies of the transfer 5-hydroxyindoleacetic acid from brain to plasma. J Pharmacol Exp Ther 158:214–218, 1967

Ninan PT, van Kammen DP, Scheinin M, et al: CSF 5-hydroxyindoleacetic acid levels in suicidal schizophrenic patients. Am J Psychiatry 141:566–569, 1984

Nyback H, Berggren BM, Nyman H, et al: Cerebroventricular volume, cerebrospinal fluid monoamine metabolites, and intellectual performance in schizophrenic patients, in Catecholamines: Neuropharmacology and Central Nervous System—Therapeutic Aspects. Edited by Usdin E, Carlsson A, Dahlstrom A. New York, Alan R Liss, 1984, pp 161–165

Partington MW, Tu JB, Wong CY: Blood serotonin levels in severe mental retardation. Dev Med Child Neurol 15:616–627, 1973

Pazos A, Probst A, Palacios JM: Serotonin receptors in the human brain. III. Autoradiographic mapping of serotonin-1 receptors. Neuroscience 21:97–122, 1987a

Pazos A, Probst A, Palacios JM: Serotonin receptors in the human brain. IV. Autoradiographic mapping of serotonin-2 receptors. Neuroscience 21:123–129, 1987b

Peroutka SJ, Snyder SH: Relationship of neuroleptic drug effects as brain dopamine, serotonin, alpha-adrenergic, and histamine receptors to clinical potency. Am J Psychiatry 137:1518–1522, 1980

Post RM, Fink E, Carpenter WT, et al: Cerebrospinal fluid amine metabolites in acute schizophrenia. Arch Gen Psychiatry 32:1063–1069, 1975

Potkin SG, Weinberger DR, Linnoila M, et al: Low CSF 5-hydroxyindoleacetic acid in schizophrenic patients with enlarged cerebral ventricles. Am J Psychiatry 140:21–25, 1983

Prosser ES, Csernansky JG, Kaplan J, et al: Depression, parkinsonian symptoms, and negative symptoms in schizophrenics treated with neuroleptics. J Nerv Ment Dis 175:100–105, 1987

Richardson BP, Engel G: The pharmacology and function of 5-HT3 receptors. Trends in Neuroscience 7:424–428, 1986

Rosengarten H, Friedhoff AJ: A review of recent studies of the biosynthesis and excretion of hallucinogens formed by methylation of neurotransmitters or related substances. Schizophr Bull 2:90–105, 1976

Scheinin M: Monoamine metabolites in human cerebrospinal fluid: indicators of neuronal activity? Med Biol 63:1–17, 1985

Sedvall G, Wode-Helgodt B: Aberrant monoamine metabolite levels in CSF and family history of schizophrenia. Arch Gen Psychiatry 37:1113–1116, 1980

Sedvall G, Fyro B, Gullberg B, et al: Relationships in healthy volunteers between concentrations of monoamine metabolites in cerebrospinal fluid and family history of psychiatric morbidity. Br J Psychiatry 136:366–374, 1980

Shore D, Korpi ER, Bigelow LB, et al: Fenfluramine and chronic schizophrenia. Biol Psychiatry 20:329–352, 1985

Sjoerdsma A, Lovenberg W, Engelman K, et al: Serotonin now: clinical implications of inhibiting its synthesis with para-chlorophenylalanine. Ann Intern Med 73:607–629, 1970

Stahl SM, Ciaranello RD, Berger PA: Platelet serotonin in schizophrenia and depression, in Serotonin in Biological Psychiatry. Edited by Ho BT, Schoolar JC, Usdin E. New York, Raven Press, 1982, pp 183–198

Stahl SM, Woo DJ, Mefford IN, et al: Hyperserotonemia and platelet serotonin uptake and release in schizophrenia and affective disorders. Am J Psychiatry 140:26–30, 1983

Stahl SM, Uhr SB, Berger PA: Pilot study on the effects of fenfluramine on negative symptoms in twelve inpatients. Biol Psychiatry 20:1098–1102, 1985

Strauss JS, Carpenter WT Jr: The prediction of outcome in schizophrenia II. Relationships between predictors and outcome variables: a report from the WHO International Pilot Study of Schizophrenia. Arch Gen Psychiatry 31:37–42, 1974

Subrahmanyam S: Role of biogenic amines in certain pathological conditions. Brain Res 87:355–362, 1975

Tyers MB: The case for 5HT$_3$ antagonists. Paper presented at the Fourth Biennial Winter Workshop on Schizophrenia, Badgastein, Austria, January 1988

van Kammen DP, Mann L, Scheinin M, et al: Spinal fluid monoamine metabolites and norepinephrine in schizophrenic patients with brain atrophy, in Pathochemical Markers in Major Psychoses. Edited by Beckman H, Reiderer P. New York, Springer-Verlag, 1985, pp 88–95

Woolley DW, Campbell NK: Exploration of the central nervous system serotonin in humans. Ann NY Acad Sci 90:108–117, 1962

Woolley DW, Shaw E: A biochemical and pharmacological suggestion about certain mental disorders. Proc Natl Acad Sci USA 40:228–231, 1954

Yesavage JA: Correlates of dangerous behavior by schizophrenics in hospital. J Psychiatr Res 18:225–231, 1984

# Chapter 10

## *Biochemical and Therapeutic Specificity of Serotonergic Agents*

William Z. Potter, M.D., Ph.D.
Matthew V. Rudorfer, M.D.

# Chapter 10

## *Biochemical and Therapeutic Specificity of Serotonergic Agents*

Most putative serotonergic drugs available for use in humans work through serotonin (5-hydroxytryptamine [5-HT]) uptake inhibition and are classified as antidepressants. The development of these drugs was based on the rationale that there was a subtype of depression that would respond preferentially to serotonergic compounds. It is currently recognized that other conditions might benefit from use of these drugs (van Praag et al. 1987), a possibility discussed in detail elsewhere in this volume. A role of serotonin in the pathophysiology or treatment of various conditions is predicated on assumptions extrapolated from preclinical research that characterizes drugs as 5-HT agonists. In order to test these underlying assumptions, it is necessary to know the extent and specificity of each drug's effects on the serotonergic system(s) in humans.

We therefore need to understand what is meant when we identify a drug as having a 5-HT–specific biochemical and/or therapeutic effect. The question of therapeutic effect is more readily approached because available data, reviewed elsewhere in this volume, support preferential treatment response to 5-HT uptake inhibitors in only one syndrome, obsessive-compulsive disorder (OCD). The critical mechanistic question is, Does this finding point to a role of 5-HT in OCD? We will argue, as van Praag et al. (1987) have done, that our current tools for studying 5-HT are not adequate for answering this or similar mechanistic questions. Nonetheless, certain evidence of biochemical specificity may emerge if a variety of parameters are simultaneously evaluated. Most of the parameters, however, are not those that are concentrated upon in preclinical studies.

## *ESTABLISHING SEROTONERGIC EFFECTS IN VIVO*

For instance, how does one best measure and interpret the acute and

233

chronic effects of 5-HT uptake inhibition? In this chapter we assume that 5-HT uptake inhibition enhances 5-HT transmission based primarily on studies in animals using in vivo electrophysiological techniques (Blier et al. 1984; de Montigny et al. 1981). Moreover, 5-HT uptake inhibition clearly can produce a "serotonergic syndrome" in rodents that have been pretreated with a monoamine oxidase inhibitor (MAOI), although the specificity of this response is questionable (Gerson and Baldessarini 1980). Alternatively, a 5-HT–releasing agent such as fenfluramine (Trulson and Jacobs 1976) or a direct 5-HT receptor agonist such as $m$-chlorophenylpiperazine (m-CPP) (Aloi et al. 1984) can be shown to produce behavioral changes in animals as well as to displace $^3$H-labeled 5-HT from binding sites in the brain and/or alter monoamine metabolism (Fuller et al. 1980). These preclinical strategies, however, cannot be directly applied in humans.

The finding in humans that most closely parallels a similar action in animals is that drug concentrations that inhibit human platelet 5-HT uptake also inhibit 5-HT uptake in rat brain slices or synaptosomes (Ross and Åberg-Wistedt 1983). Acute effects of serotonergic drugs in humans have also been assessed in terms of neuroendocrine responses, specifically plasma prolactin and cortisol, to central serotonergic probes. There is much less preclinical work with this approach, although in rats pretreatment with 5-HT uptake inhibitors enhances the prolactin response to 5-HT precursors such as 5-hydroxytryptophan (Meltzer et al. 1981). Furthermore, m-CPP has been shown to stimulate prolactin release in rats and monkeys (Aloi et al. 1984; Quattrone et al. 1981).

The advantage of hormonal responses is that they can be measured in the blood. The seemingly obvious measures of 5-HT and its major metabolite, 5-hydroxyindoleacetic acid (5-HIAA), in plasma and/or urine are possible but have not been extensively pursued. This is because the bulk of 5-HT in the blood is believed to be dissociated from CNS function. The concentration of 5-HT is highest in the platelets, which actively concentrate 5-HT for possible utilization in steps subsequent to aggregation; the 5-HT apparently is principally formed in the enterochromaffin cells of the intestine. Whatever the source of the 5-HT in peripheral tissues, its storage and metabolism in these areas may have parallels to its storage and metabolism in the brain. Thus documentation of the behavior of 5-HT in any compartment may be useful.

Much greater attention, however, has been devoted to assessing alterations in the CSF concentration of the 5-HT metabolite 5-HIAA. Using each subject as his or her own control, if a drug does not alter clearance of 5-HIAA from the CSF or alter the fraction of 5-HT metabolized to 5-HIAA, then its effects on 5-HIAA may be interpreted as being proportional to its effects on 5-HT turnover.

## ARE SEROTONERGIC EFFECTS INDEPENDENT OF EFFECTS ON OTHER SYSTEMS?

Because specificity is the focus, it is necessary to establish the extent to which other neurotransmitter systems are affected. Endocrine responses can be evoked by a large variety of agents, including those that work through adrenergic, dopaminergic, GABA-ergic, and, most obviously, peptidergic receptors. To establish serotonergic specificity, the most feasible approach in humans is to pretreat with a serotonergic antagonist and show that it blocks a response. However, the characterization of 5-HT receptors is still evolving without clear relationships having yet been established between receptor subtypes and specific functions (Conn and Sanders-Bush 1987). In none of the studies discussed below has a sufficiently wide range of antagonists been used to convincingly identify the 5-HT receptor subtype(s) that is responsible for the release of either prolactin or cortisol in humans.

The other control for specificity is demonstration that other neurotransmitters either are not affected following a serotonergic stimulus or are affected in a way that would reflect predictable interactions between, for instance, 5-HT and dopamine (Ågren et al. 1986). There has been increasing attention paid to functional neurotransmitter interactions in preclinical studies (Manier et al. 1987; Martin-Iverson et al. 1983; Oades 1985), but it is far too early to extrapolate these findings to humans. What we can do is concurrently measure indices of other neurotransmitters, which include the dopamine metabolite homovanillic acid (HVA) in CSF, and norepinephrine or its major metabolites, especially 3-methoxy-4-hydroxyphenylglycol (MHPG), in CSF, plasma, and urine. To our knowledge there are no studies on neurotransmitter measures in human CSF following acute administration of newer putative seotonergic drugs; for all practical purposes, one is therefore limited to measures of norepinephrine and MHPG in plasma. Even these measures are rarely checked in acute stimulation paradigms. Following chronic administration of serotonergic drugs, however, sufficient attention has been paid to measures in CSF, plasma, and urine to show that changes frequently (and perhaps inevitably) occur in dopaminergic and/or noradrenergic measures, rendering it virtually impossible to assert that any chronic effect is directly the effect of serotonergic stimulation (Potter et al. 1985).

It is also relevant to recall that until very recently many of the agents developed as serotonergic drugs have had considerable inherent lack of specificity even in animal models. Some 5-HT uptake inhibitors are metabolized to active metabolites that have a different pharmacodynamic profile. One such inhibitor, clomipramine, produces higher plasma concentrations of its main desmethylated metabolite than of clomipramine itself (Träskman et al. 1979). This metabolite is a relatively potent inhibitor of norepinephrine uptake; hence, chronic orally administered clomipramine cannot be claimed to be primarily serotonergic. The relative potencies of parent drug and known active metabolites for 5-HT and norepinephrine uptake for a series of compounds are shown in Table 10-1. On the basis of available evidence it would appear that citalopram, fluoxetine, and perhaps fluvoxamine (for which there are few data on metabolites in humans)

**Table 10-1.** Parent drugs and metabolites that inhibit monoamine uptake: relative potency for serotonin versus norepinephrine uptake

| Parent drug | $IC_{50}$–NE/5-HT | Active metabolite | $IC_{50}$–NE/5-HT |
|---|---|---|---|
| First-generation tricyclic antidepressants | | | |
| Amitriptyline | 1.0 | Nortriptyline | 0.05 |
| Clomipramine | 10 | Norclomipramine | 0.3 |
| Second-generation antidepressants with identified uptake-inhibiting metabolite | | | |
| Zimelidine | 17 | Norzimelidine | 10 |
| Fluoxetine | 24 | Norfluoxetine | 19 |
| Citalopram | 733 | Norcitalopram | >700 |
| Second-generation antidepressants without known uptake-inhibiting metabolites | | | |
| Trazodone | 15 | | |
| Femoxetine | 20 | | |
| Fluvoxamine | 41 | | |
| Indalpine | 50 | | |
| Paroxetine | 320 | | |

*Note.* Ratio reflects $IC_{50}$ (μm) for norepinephrine (NE) uptake divided by $IC_{50}$ (μm) for serotonin (5-HT) uptake in rat brain synaptosomes. The higher the ratio, the greater the relative specificity for 5-HT. Values taken from Hall and Ogren 1981; Hyttel 1982; Richelson 1979; Wong et al. 1983.

are the most likely to retain their selectivity for 5-HT uptake under actual clinical conditions (Bjerkenstedt et al. 1985a; Stark et al. 1985). Other 5-HT uptake inhibitors such as paroxetine, femoxetine, and indalpine may also prove "clean" in this regard (Buus Lassen et al. 1975; LeFur and Uzan 1977; Reebye et al. 1982; Thomas et al. 1987). Trazodone is relatively specific for 5-HT uptake but is also relatively weak as an inhibitor (Stark et al. 1985) and is therefore not here considered as part of the group under discussion.

With regard to fenfluramine, the single other type of agent most researched as a 5-HT drug, it has long been clear from preclinical studies that, depending on the dose and species, both norepinephrine and dopamine releases can be affected (Calderini et al. 1975; Orosco et al. 1984; Rowland and Carlton 1986). Direct 5-HT agonists such as m-CPP may emerge as being far more specific, although they have the inherent disadvantage of being less physiologic, because exogenous direct agonists act on receptors whatever their localization. Physiologically released or increased 5-HT will act on 5-HT receptors following a more restricted distribution, i.e., that conforming to the distribution of serotonergic neurons rather than to the distribution of receptor sites. Autoradiographic studies point to many areas of "mismatch" between the distribution of serotonergic neurons and receptors (de Souza and Kuyatt 1987; Pazos et al. 1985; Rainbow and Biegon 1983).

Detailed descriptions of the acute effects of various serotonergic agents on neuroendocrine, physiologic, and behavioral parameters are provided elsewhere in this volume. We will therefore focus on reviewing chronic biochemical effects of those drugs administered as potential therapeutic agents for depression, OCD, and in the case of fenfluramine, for autism or hyperactivity. A few comments, however, about the acute effects of the newer 5-HT uptake inhibitors will be useful to remind us that these agents do represent an advance in specificity.

## ACUTE BIOCHEMICAL EFFECTS OF SEROTONIN UPTAKE INHIBITORS

Ideally, a selective 5-HT uptake inhibitor would have that inhibition as its single biochemical effect. All of the so-called first-generation antidepressants (Table 10-1) not only inhibit 5-HT and/or norepinephrine uptake, but also block muscarinic, histaminergic, and $\alpha_1$ receptors to a sufficient extent that effects secondary to such blockade can be measured in humans (Richelson 1979). The second-generation 5-HT uptake inhibitors are generally free of these actions (Hall and Ogren 1981; Hyttel 1982; Wong et al. 1983). To our knowledge there is no study in humans showing significant antimuscarinic, antihistaminergic, or $\alpha_1$-blocking activity after administration of zimelidine, fluvoxamine, citalopram, fluoxetine, paroxetine, or femoxetine. The most recent compounds have not, however, been exhaustively evaluated in this regard, so some such effects could emerge. Moreover, 5-HT uptake inhibitors are not acutely sedating. This characteristic stands in contrast to the sedation produced by most tricyclic antidepressants presumably through their blockade of histamine and/or $\alpha_1$ receptors (Richelson 1979).

Indeed, acute oral administration of these drugs produces few measurable effects beyond inhibition of 5-HT uptake by platelets. In doses up to 200 mg, orally administered zimelidine did not alter plasma prolactin or growth hormone levels; nor did it affect blood pressure, heart rate, or plasma levels of catecholamines (Calil et al. 1984). Furthermore, there is no available evidence to support that orally administered fluvoxamine, citalopram, or fluoxetine will acutely alter these parameters. Interestingly, there may be a weak stimulating effect of orally administered zimelidine on plasma cortisol levels, and in the case of fluoxetine, single oral doses of 30 mg clearly double plasma cortisol levels (Petraglia et al. 1984). Intravenous administration of 5-HT uptake inhibitors should increase levels of prolactin, as well as ACTH and cortisol, if extrapolations from rodents administered fluoxetine (Stark et al. 1985) or citalopram (Lesieur et al. 1985) are predictive. The mechanism of the prolactin level increase appears to be complex, because it may involve a 5-HT-mediated inhibition of dopamine release (Pilotte and Porter 1981). Nonetheless, studies of prolactin, as well as cortisol, release have been pursued as a measure of serotonergic function in humans.

It was shown several years ago in man that intravenously administered clomipramine can transiently increase prolactin levels at doses that minimally affect growth hormone levels (Laakman et al. 1984). We have replicated this finding and reduced the dose of clomipramine to such an extent (10–12 mg/subject) that there is no effect on growth hormone levels, but an increased level of prolactin, cortisol, and ACTH (Golden et al. 1989). Under these conditions, the desmethyl metabolite of clomipramine does not appear as a confound; it is undetectable at the time of peak hormone response.

Because at least one 5-HT uptake inhibitor, fluoxetine, has been shown in the rat to increase release of corticotropin-releasing factor into hypophyseal portal blood (Gibbes and Vale 1983), ACTH and/or cortisol responses are at least equally relevant. The data suggest that observable acute neuroendocrine effects of 5-HT uptake inhibitors are probably caused by an acute increase of 5-HT operating at relevant sites. None of the changes suggest that these drugs are having other primary effects. As for the two other classes of drugs— 5-HT–releasing agents, as exemplified by fenfluramine, and direct agonists, as exemplified by m-CPP (Mueller et al. 1985)—both preclinical evidence and clinical evidence presented elsewhere in this volume support the interpretation that the acute primary effect of both classes of drugs in doses used in humans is on components of the serotonergic system. However, at least in the rat, m-CPP produces other effects, such as increased plasma catecholamine levels, through possible central serotonergic mechanisms (Bagdy et al. 1988).

## CHRONIC BIOCHEMICAL EFFECTS OF SEROTONIN UPTAKE INHIBITORS AND FENFLURAMINE

Following chronic administration, the newer 5-HT uptake inhibitors behave as predicted in some respects, apparently having no effects on cardiovascular function and salivary flow (Christensen et al. 1985; Rafaelsen et al. 1981). Not predicted, however, was the apparent absence of effects on plasma prolactin or cortisol during chronic administration (Calil et al. 1984). The lack of such changes presumably reflects homeostatic mechanisms that effectively maintain balanced neuroendocrine output in the face of brain monoamine neurotransmitter alterations.

As expected, chronic treatment produces continued inhibition of 5-HT uptake in platelets, an effect that can be directly related to plasma concentrations. For instance, after 4 weeks of citalopram administration with weekly blood samples taken, there was a .96 correlation between drug concentration and inhibition of 5-HT by platelets (Bjerkenstedt et al. 1985b). Thus whatever compensatory changes may occur "downstream," the primary effect of 5-HT uptake inhibition persists. Similarly, after chronic administration of fenfluramine, evidence of a sustained 5-HT–releasing effect can be seen in terms of depletion of platelet or whole-blood 5-HT (Donnelly et al. 1989; Ho et al. 1986). On the other hand, one also observes evidence of norepinephrine depletion by fenfluramine, as both plasma and/or urinary norepinephrine and MHPG levels are reduced (de la Vega et al. 1977; Donnelly et al. 1989). These findings bring us to a consideration of whether chronic effects on other neurotransmitters should be interpreted as evidence of nonspecificity or of secondary compensatory changes resulting from a primary serotonergic effect.

In the introduction to this chapter we identified monoamine measures in the CSF as providing measures of at least 5-HT and dopamine formation and metabolism. Reductions in CSF concentrations of 5-HIAA, the major 5-HT metabolite, are hypothesized to result from chronically increased intrasynaptic 5-HT. Interestingly, all tricyclic antidepressants, even the putatively norepinephrine-selective desipramine, reduce concentrations of 5-HIAA in the CSF following chronic administration (Figure 10-1; Table 10-2). An early study reports a similar effect even after fenfluramine administration (Shoulson and Chase 1975). As expected, the second-generation 5-HT uptake inhibitors also consistently reduce 5-HIAA levels. Surprisingly, at least in depressed patients, zimelidine and citalopram also reduce levels of the norepinephrine metabolite MHPG in CSF (Figure 10-1) (Bertilsson et al. 1980; Bjerkenstedt et al. 1985a; Potter et al. 1985).

It should also be noted that 5-HT uptake inhibitors can be distinguished from tricyclic antidepressants in terms of a tendency to increase levels of homovanillic acid (HVA), which are variably and inconsistently affected by the first-generation compounds (Table 10-3; Figure 10-1). It is important to consider possible interpretations for the common tendency of both classic tricyclic antidepressants and nontricyclic second-generation 5-HT uptake inhibitors to reduce levels of 5-HIAA and MHPG while the two drug classes have different effects on HVA.

Interestingly, fenfluramine has been reported to decrease markedly 5-HIAA levels and to tend to decrease HVA levels in CSF (Meyen-

**Table 10-2.**   Treatment-associated decrease in 5-hydroxyindoleacetic acid in human cerebrospinal fluid: percent reduction from pretreatment value in American and Scandinavian studies

| Treatment | Mean percent reduction (%) |
|---|---|
| Imipramine | 21, 22 |
| Desipramine | 32 |
| Amitriptyline | 21, 37 |
| Nortriptyline | 10, 25 |
| Clomipramine | 44, 47 |
| Zimelidine | 20, 35, 39 |
| Citalopram | 29 |
| Fluvoxamine | 24 |

*Note.* Data taken from Åsberg et al. 1973; Bertilsson et al. 1974, 1980; Bjerkenstedt et al. 1985a; Bowden et al. 1985; Martin et al. 1987; Potter et al. 1985; Träskman et al. 1979.

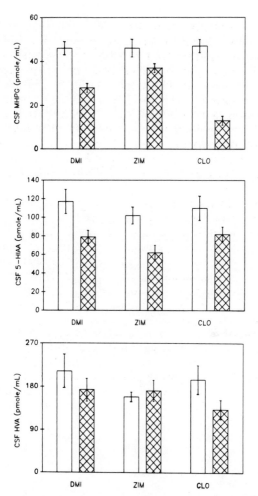

**Figure 10-1.** Comparative drug effects on CSF monoamine metabolite concentrations during chronic treatment of depressed patients. Data ($\bar{x} \pm$ SEM) are shown for a norepinephrine reuptake inhibitor, desipramine (DMI, $n = 11$), a serotonin reuptake inhibitor, zimelidine (ZIM, $n = 9$), and a monoamine oxidase type A inhibitor, clorgyline (CLO, $n = 10$). All effects are significant at the level of $P < .02$ except for changes in homovanillic acid (HVA) with DMI or ZIM, which are not significant. MHPG = 3-methoxy-4-hydroxyphenylglycol; 5-HIAA = 5-hydroxyindoleacetic acid. (Drawn from data in Potter et al. 1985.)

dorff et al. 1986). Indeed, fenfluramine appears to have a general monoamine-depleting action such that urinary excretion of epinephrine, norepinephrine, and MHPG is reduced by 48%, 67%, and 33%, respectively, while platelet 5-HT content is reduced by 36%. Moreover, fenfluramine produces a modest but significant reduction of blood pressure (Donnelly et al. 1989). Such a spectrum of effects is suggestive of a direct reserpine-like action, suggesting that many effects of fenfluramine are neither directly nor indirectly related to its effects on 5-HT.

## SECONDARY INTERACTIONS OF MONOAMINE SYSTEMS

First, as we and others have previously discussed (Frances et al. 1987; Manier et al. 1987; Potter et al. 1985), there is an increasing body of data showing neuroanatomical and functional links between the noradrenergic and serotonergic systems in the CNS such that drugs affecting one are likely to affect the other. As discussed below, we do not believe that the common reductions of MHPG and 5-HIAA levels reflect nonspecificity in terms of uptake inhibition, because the pat-

**Table 10-3.**  Effects of first-generation tricyclic antidepressants versus second-generation serotonin uptake inhibitors on homovanillic acid in human cerebrospinal fluid

|  | Percent change in HVA levels | Statistical significance |
| --- | --- | --- |
| First-generation tricyclic antidepressants |  |  |
| Imipramine | +7 | NS |
| Desipramine | −19 | NS |
| Amitriptyline | −3 | NS |
| Nortriptyline | −10 | NS |
| Clomipramine | +11 | NS |
| Second-generation serotonin uptake inhibitors |  |  |
| Zimelidine | +20 | $P < .01$ |
| Citalopram | +15 | $P < .05$ |

*Note.* NS = Not significant; HVA = homovanillic acid. Data taken from Ågren et al. 1988; Åsberg et al. 1977; Bertilsson et al. 1980; Bjerkenstedt et al. 1985a; Bowden et al. 1985; Potter et al. 1985.

tern of metabolism of norepinephrine is predictably altered in the direction of a relative increase in the extraneuronal metabolite normetanephrine by norepinephrine and not by 5-HT uptake inhibitors. Thus, the reduction of MHPG after zimelidine administration in both depressed patients and nondepressed volunteers (Rudorfer et al. 1984) and after citalopram administration, at least in depressed patients, could result from a serotonergic enhancement of aspects of norepinephrine function that produce a compensatory decrease in norepinephrine turnover.

The tendency of 5-HT uptake inhibitors to increase HVA levels in CSF may be explained, in turn, by another type of interaction—that of the serotonergic and dopaminergic systems. Briefly, all carefully done CSF studies employing accepted methods of quantitation reveal a high positive correlation between 5-HIAA and HVA in the CSF. Moreover, drug-induced changes in levels of HVA and 5-HIAA are highly correlated, a point noted in previous studies of clomipramine and citalopram (Åsberg et al. 1977; Bjerkenstedt et al. 1985a). Either within most brain regions or across brain regions in animals, there is a high correlation between 5-HT and dopamine as well as between 5-HIAA and HVA. Electrophysiological and behavioral studies in animals demonstrate functional relationships between the serotonergic and dopaminergic systems. Finally, 5-HT neurons in the dorsal raphe nuclei project directly to major dopamine-containing areas such as the substantia nigra. We have interpreted this body of data as being most consistent with a positive influence of 5-HT turnover on dopamine turnover (Ågren et al. 1986).

In the future it will be of interest to investigate the chronic biochemical effects of direct agonists such as m-CPP and of antagonists such as ritanserin. A report on selected effects of chronic ritanserin reveals the expected persisting blockade of 5-HT–agonist binding to platelets but also (to the authors) a curious failure to alter L-tryptophan–induced prolactin release (Idzikowski et al. 1987). We suggest that a more extensive pharmacodynamic profile of such drugs is needed before forming expectations. Such a profile could include parameters in plasma and urine that, at least in the case of the norepinephrine system, can be highly interrelated to central measures (Maas and Leckman 1983).

## IMPORTANCE OF URINARY MEASURES IN ASSESSING MONOAMINERGIC EFFECTS OF DRUGS: OUTPUT VERSUS METABOLISM

There are special advantages to urinary measures in trying to establish specificity of a particular drug. It is possible to obtain integrated measures of norepinephrine and its metabolites in urine over time, in contrast to the one or two time points available with CSF; thus, this peripheral measure can address the question of whether a drug alters turnover, pathways of metabolism, or both. We have employed the sum of norepinephrine and its major metabolites in sequential 24-hour urine samples as a measure of the "whole-body" turnover of norepinephrine. In the same sample one can look at the relative amount of the total urine that is excreted as norepinephrine, MHPG, vanillylmandelic acid (VMA), or normetanephrine. A norepinephrine uptake inhibitor such as desipramine both decreases turnover (the sum) and alters the relative metabolism, since the fractional excretion of the extraneuronally formed normetanephrine actually increases. Zimelidine, however, reduces total turnover without affecting the proportional excretion of specific metabolites; thus, zimelidine does not appear to be acting as a norepinephrine uptake inhibitor like desipramine, although it shares the property of reducing whole-body norepinephrine turnover (Linnoila et al. 1982; Rudorfer et al. 1984).

Interestingly, we could not show an effect of zimelidine on urinary 5-HT and 5-HIAA output (Linnoila et al. 1984). This suggests that 5-HT uptake inhibitors can act on the CNS so as to produce modest secondary reduction of peripheral sympathetic outflow. To our knowledge, the studies of urinary output after administration of 5-HT uptake inhibitors other than zimelidine that are necessary to support this generalization have not been performed. It will be of interest to see if drugs such as fluoxetine and fluvoxamine reduce whole-body norepinephrine turnover without affecting total 5-HT or 5-HIAA output in urine.

## THE NEED TO SHOW SPECIFICITY OF COMPOUNDS AS PATTERN OF EFFECTS

It should be obvious from the above that those drugs that we label as serotonergic have a variety of biochemical effects, especially following chronic administration. Any "specificity" must be shown for each class of compounds and would appear to refer more to some unique *pattern* of effects rather than to some pure isolated serotonergic alteration. It does seem reasonable to conclude that the agents that we have discussed are substantively distinct in the ways described. Whether they provide adequate tools to explore possible pathophysiologic roles of serotonergic abnormalities is doubtful. This caution is reinforced when we turn to a brief consideration of clinical specificity.

## CLINICAL SPECIFICITY: TARGET PATHOLOGY VERSUS SIDE EFFECTS

Clinical specificity of a drug generally refers to efficacy in a particular subtype of an illness. In the case of the 5-HT reuptake inhibitors, this would require a well-defined serotonergic-dysfunction state. As mentioned, the one disorder in which 5-HT has been implicated is OCD (reviewed in this volume by Zohar et al., Chapter 5). Drugs that work in OCD are 5-HT uptake inhibitors—clomipramine, fluoxetine (Fontaine and Chouinard 1986; Turner et al. 1985), and fluvoxamine (Perse et al. 1987; Price et al. 1987).

In other conditions, however, no such associations between biochemical specificity and response emerge. Descriptions in some samples of a subset of depressed patients exhibiting low CSF 5-HIAA concentrations who may have responded to a more serotonergic drug (Åsberg et al. 1973) have generally not been replicated (reviewed by Davis and Bresnahan 1987). Clinically, the selective 5-HT reuptake inhibitors, like the less-selective antidepressants, exhibit a broad range of activity in major depression. Nonetheless, their lack of sedation and, in some patients, their activating effects do offer advantages.

Conversely, any stimulatory effects (which may or may not reflect 5-HT uptake inhibition) may confound tests of therapeutic specificity in other conditions such as panic disorder patients. In one recent open study (Gorman et al. 1987), patients treated with fluoxetine reported a favorable response in terms of panic attacks, but fully half of the patients were unable to tolerate the increased agitation, jitteriness, and associated symptoms of this medication. As with virtually all effective antidepressant treatments, individuals on fluoxetine may become manic (Chouinard and Steiner 1986; Settle and Settle 1984; Turner et al. 1985), an observation that seems to argue against specificity. The *lack* of certain side effects, however, is an advantage in the search for clinical specificity. For instance, like zimelidine before it, fluoxetine does not potentiate (or block) the effects of ethanol (Lemberger et al. 1985). As noted, the lack of affinity of the selective 5-HT reuptake inhibitors for a variety of receptors in vitro, including muscarinic cholinergic, α-adrenergic, and histaminergic receptors, translates clinically into a lack of anticholinergic, hypotensive, and sedative side effects. These compounds also are not associated with typical tricyclic antidepressant–like prolongation of cardiac condition time on the ECG.

However, the selective 5-HT inhibitors can produce "nonspecific" side effects, including nervousness and anxiety, insomnia, gastrointestinal disturbances (nausea, diarrhea), headache, and anorexia (Somni et al. 1987; Wernicke 1985). An exaggerated form of these toxicities, resembling the "serotonin syndrome" in animal models, has been described in patients receiving a combination of fluoxetine and L-tryptophan (Steiner and Fontaine 1986). If some effects of fluoxetine are 5-HT–specific, such as the reported anorexic effect (Ferguson 1986; Nelson and Pool 1987), then such drugs could prove useful in studying eating disorders.

This brings us back to a consideration of specific psychotherapeutic uses of fenfluramine, a drug marketed as an anorectic. This drug has been tried, without success, in hyperactivity (Donnelly et al. 1989). In autism, however, preliminary studies suggested that as many as one-third of patients would improve (Ritvo et al. 1986). This finding superficially at least made theoretical sense, because autistic children are reported to have elevated 5-HT levels in blood, which fenfluramine can be shown to reduce (Ho et al. 1986). Initial studies, however, appear to have been methodologically flawed, and more recent placebo-controlled studies using a randomized order of drug presentation have not revealed a beneficial effect of fenfluramine in autism (Campbell 1988). (See Chapter 3, this volume, for discussion of this issue.) Other possible therapeutic roles of serotonergic drugs are discussed elsewhere in this volume, although specificity does not appear to be a central issue, because these roles involve syndromes such as anxiety and eating disorders in which other classes of agents are already extensively used.

## CONCLUSIONS

Without well-understood serotonergic probes and well-defined relationships between 5-HT agonists, receptors, and effects, and given the complex interrelationships between 5-HT and other neurotransmitters, no simple pure 5-HT–dependent syndrome or drug response is likely to be identified. Nonetheless, at least selective 5-HT uptake inhibitors do yield unique patterns of biochemical and physiological effects. To this extent, the pharmacologic data support the strategy of using selective response to these drugs as a tool for exploring the pathophysiology of a syndrome. Currently, the one condition for which treatment appears to require primary serotonergic stimulation is OCD. We suggest that it will be through studies on this syndrome that the biochemical and therapeutic specificity of serotonergic agents may be most productively explored in the near future.

## REFERENCES

Ågren H, Mefford I, Rudorfer M, et al: Interacting neurotransmitter systems: a non-experimental approach to the 5HIAA-HVA correlation in human CSF. J Psychiatr Res 20:175–193, 1986

Ågren H, Nordin C, Potter WZ: Antidepressant drug action and CSF monoamine metabolites, in Progress in Catecholamine Research, Part C: Clinical Aspects. Edited by Belmaker RH, Sandler M, Dahlstrom A, et al. New York, Alan R Liss, 1988, pp 307–311

Aloi JA, Insel TR, Mueller EA, et al: Neuroendocrine and behavioral effects of *m*-chlorophenylpiperazine administration in rhesus monkeys. Life Sci 34:1325–1331, 1984

Åsberg M, Bertilsson L, Tuck JR, et al: Indoleaminic metabolites in the cerebrospinal fluid of depressed patients before and during treatment with nortriptyline. Clin Pharmacol Ther 14:277–286, 1973

Åsberg M, Ringberger V-A, Sjöqvist F, et al: Monoamine metabolites in cerebrospinal fluid and serotonin uptake inhibition during treatment with chlorimipramine. Clin Pharmacol Ther 21:201–207, 1977

Bagdy G, Szemeredi K, Hill JL, et al: The serotonin agonist, *m*-chlorophenyl-piperazine, markedly increases levels of plasma catecholamines in the conscious rat. Neuropharmacology 10:975–980, 1988

Bertilsson L, Åsberg M, Thorén P: Differential effect of chlorimipramine and nortriptyline on cerebrospinal fluid metabolites of serotonin and noradrenaline in depression. Eur J Clin Pharmacol 7:365–368, 1974

Bertilsson L, Tuck JR, Siwers B: Biochemical effects of zimelidine in man. Eur J Clin Pharmacol 18:483–487, 1980

Bjerkenstedt L, Edman G, Flyckt L, et al: Clinical and biochemical effects of citalopram, a selective 5-HT reuptake inhibitor—a dose-response study in depressed patients. Psychopharmacology (Berlin) 87:253–259, 1985a

Bjerkenstedt L, Flyckt L, Fredericson Overo K, et al: Relationship between clinical effects, serum drug concentration and serotonin uptake inhibition in depressed patients treated with citalopram: a double-blind comparison of three dose levels. Eur J Clin Pharmacol 28:553–557, 1985b

Blier P, de Montigny C, Tardif D: Effects of the two antidepressant drugs mianserin and indalpine on the serotonergic system: single-cell studies in the rat. Psychopharmacology (Berlin) 84:242–249, 1984

Bowden CL, Koslow SH, Hanin I, et al: Effects of amitriptyline and imipramine on brain amine neurotransmitter metabolites in cerebrospinal fluid. Clin Pharmacol Ther 37:316–324, 1985

Buus Lassen J, Petersen E, Kjellberg B, et al: Comparative studies of a new 5HT-uptake inhibitor and some tricyclic thymoleptics. Eur J Pharmacol 32:108–115, 1975

Calderini G, Morselli P, Garattini S: Effect of amphetamine and fenfluramine on brain noradrenaline and MOPEG-SO4. Eur J Pharmacol 34:345–350, 1975

Calil HM, Lesieur P, Gold PW, et al: Hormonal responses to zimelidine and desipramine in depressed patients. Psychiatry Res 13:231–242, 1984

Campbell M: Annotation: fenfluramine treatment of autism. J Child Psychol Psychiatry 29:1–10, 1988

Chouinard G, Steiner W: A case of mania induced by high-dose fluoxetine treatment. Am J Psychiatry 143:686, 1986

Christensen P, Thorsen HY, Pedersen OL, et al: Orthostatic side effects of clomipramine and citalopram during treatment for depression. Psychopharmacology (Berlin) 86:383–385, 1985

Conn PJ, Sanders-Bush E: Central serotonin receptors: effector systems, physiological roles and regulation. Psychopharmacology (Berlin) 92:267–277, 1987

Davis JM, Bresnahan DB: Psychopharmacology in clinical psychiatry, in Psychiatry Update: American Psychiatric Association Annual Review, Vol 6. Edited by Hales RE, Frances AJ. Washington, DC, American Psychiatric Press, 1987, pp 159–187

de la Vega CE, Slater S, Ziegler MG, et al: Reduction in plasma norepinephrine during fenfluramine treatment. Clin Pharmacol Ther 21:216–221, 1977

de Montigny C, Blier P, Caille G, et al: Pre- and post-synaptic effects of zimelidine and norzimelidine on the serotonergic system: single cell studies in the rat. Acta Psychiatr Scand 63 (suppl 290):79–90, 1981

de Souza EB, Kuyatt BI: Autoradiographic localization of $^{3}$H-paroxetine-labeled serotonin uptake sites and rat brain. Synapse 1:488–496, 1987

Donnelly M, Rapoport JL, Potter WZ, et al: Fenfluramine and amphetamine treatment of childhood hyperactivity: clinical and biochemical findings. Arch Gen Psychiatry 46:205–212, 1989

Feighner J, Hendrickson G, Miller L, et al: Double-blind comparison of doxepin versus bupropion in outpatients with a major depressive disorder. J Clin Psychopharmacol 6:27–32, 1986

Ferguson JM: Fluoxetine-induced weight loss in overweight, nondepressed subjects. Am J Psychiatry 143:1496, 1986

Fontaine R, Chouinard G: An open clinical trial of fluoxetine in the treatment of obsessive-compulsive disorder. J Clin Psychopharmacol 6:98–101, 1986

Frances H, Raisonan R, Simon P, et al: Lesions of the serotonergic system impair the facilitation of but not the tolerance to the effects of chronic clenbuterol administration. Psychopharmacology (Berlin) 91:496–499, 1987

Fuller RW, Mason NR, Molloy BB: Structural relationships in the inhibition of [$^3$H]-serotonin binding to rat brain membranes in vitro by 1-phenyl-piperazines. Biochem Pharmacol 29:833–835, 1980

Gerson SC, Baldessarini RJ: Minireview: motor effects of serotonin in the central nervous system. Life Sci 27:1435–1451, 1980

Gibbes DM, Vale W: Effect of the serotonin reuptake inhibitor fluoxetine on corticotropin-releasing factor and vasopressin secretion into hypophyseal portal blood. Brain Res 280:176–179, 1983

Golden RN, Hsiao J, Lane E, et al: The effects of intravenous clomipramine on neurohormones in normal subjects. J Clin Endocrinol Metab 68:632–37, 1989

Gorman JM, Liebowitz MR, Fyer AJ, et al: An open trial of fluoxetine in the treatment of panic attacks. J Clin Psychopharmacol 7:329–332, 1987

Hall H, Ogren S-O: Effects of antidepressant drugs on different receptors in the brain. Eur J Pharmacol 70:393–407, 1981

Ho HH, Lockitch G, Eaves L, et al: Blood serotonin concentrations and fenfluramine therapy in autistic children. J Pediatr 108:465–469, 1986

Hyttel J: Citalopram-pharmacological profile of a specific serotonin uptake inhibitor with antidepressant activity. Prog Neuropsychopharmacol Biol Psychiatry 6:277–295, 1982

Idzikowski C, Cowen PJ, Nutt D, et al: The effects of chronic ritanserin treatment on sleep and the neuroendocrine response to L-tryptophan. Naunyn Schmiedebergs Arch Pharmacol 335:109–114, 1987

Laakman G, Gugath M, Kuss HJ, et al: Comparison of growth hormone and prolactin stimulation induced by chlorimipramine and desipramine in man in connection with chlorimipramine metabolism. Psychopharmacology (Berlin) 82:62–67, 1984

LeFur G, Uzan A: Effects of 4(3-indolyl-alkyl) piperidine derivatives on uptake and release of noradrenaline, dopamine and 5-hydroxytryptamine. Biochem Pharmacol 26:479–503, 1977

Lemberger L, Rowe H, Bergstrom RF, et al: Effect of fluoxetine on psychomotor performance, physiologic response, and kinetics of ethanol. Clin Pharmacol Ther 37:658–664, 1985

Lesieur P, Saavedra JM, Chiueh CC, et al: Serotonin versus dopamine in prolactin release. Paper presented at the annual meeting of the Society of Biological Psychiatry, Dallas, TX, May 15–19, 1985

Linnoila M, Karoum F, Calil HM, et al: Alteration of norepinephrine metabolism with desipramine and zimelidine in depressed patients. Arch Gen Psychiatry 39:1025–1028, 1982

Linnoila M, Miller T, Bartko J, et al: Five antidepressant treatments in depressed patients: effects on urinary serotonin and 5-hydroxyin-doleacetic acid output. Arch Gen Psychiatry 41:688–692, 1984

Maas JW, Leckman JF: Relationships between central nervous system noradrenergic function and plasma and urinary MHPG and other norepinephrine metabolites, in MHPG: Basic Mechanisms and Psychopathology. Edited by Maas JW. New York, Academic Press, 1983, pp 33–43

Maas JW, Koslow SH, Katz MM, et al: Pretreatment neurotransmitter metabolites and tricyclic antidepressant drug response. Am J Psychiatry 141:1159–1171, 1984

Manier DH, Gillespie DD, Sanders-Bush E, et al: The serotonin/-noradrenaline link in brain. I. The role of noradrenaline and serotonin in the regulation of density and function of beta adrenoceptors and its alteration by desipramine. Naunyn Schmiedebergs Arch Pharmacol 335:109–114, 1987

Martin PR, Adinoff B, Bone GAH, et al: Fluvoxamine treatment of alcoholic chronic organic brain syndromes. Clin Pharmacol Ther 41:211, 1987

Martin-Iverson MT, Leclere J-F, Fibiger HC: Cholinergic-dopaminergic interactions and the mechanisms of action of antidepressants. Eur J Pharmacol 94:193–201, 1983

Meltzer HY, Simonovic M, Sturgeon RD, et al: Effects of antidepressants, lithium and electroconvulsive treatment on rat serum prolactin levels. Acta Psychiatr Scand 63 (suppl 290):100–126, 1981

Meyendorff E, Jain A, Träskman-Bendz L, et al: The effects of fenfluramine on suicidal behavior. Psychopharmacol Bull 22:155–159, 1986

Mueller EA, Murphy DL, Sunderland T: Neuroendocrine effects of m-chlorophenylpiperazine, a serotonin agonist, in humans. J Clin Endocrinol Metab 61:1179–1184, 1985

Nelson EB, Pool JL: Fluoxetine in the treatment of obesity. Clin Pharmacol Ther 41:198, 1987

Oades RD: The role of noradrenaline in tuning and dopamine in switching between signals in the CNS. Neurosci Biobehav Rev 9:261–282, 1985

Orosco M, Bremond J, Jaquot D, et al: Fenfluramine and brain transmitters in the obese rat. Neuropharmacology 23:183–188, 1984

Pazos A, Cortes R, Palacios JM: Quantitative autoradiographic mapping of serotonin receptors in the rat brain. II. Serotonin-2 receptors. Brain Res 346:231–249, 1985

Perse TL, Greist JH, Jefferson JW, et al: Fluvoxamine treatment of obsessive-compulsive disorder. Am J Psychiatry 144:1543–1548, 1987

Petraglia F, Facchinetti F, Martignoni E, et al: Serotoninergic agonists increase plasma levels of β-endorphin and β-lipotropin in humans. J Clin Endocrinol Metab 59:1138–1142, 1984

Pilotte NW, Porter JC: Dopamine in hypophyseal portal plasma and prolactin in systemic plasma of rats treated with 5-hydroxytryptamine. Endocrinology 108:2137–2141, 1981

Potter WZ: Psychotherapeutic drugs and biogenic amines: current concepts and therapeutic implications. Drugs 28:127–143, 1984

Potter WZ, Scheinin M, Golden RN, et al: Selective antidepressants lack specificity on norepinephrine and serotonin metabolites in cerebrospinal fluid. Arch Gen Psychiatry 42:1171–1177, 1985

Price LH, Goodman WK, Charney DS, et al: Treatment of severe obsessive-compulsive disorder with fluvoxamine. Am J Psychiatry 144:1059–1061, 1987

Quattrone A, Schettini G, Anunziato L, et al: Pharmacological evidence of super-sensitivity of central serotonergic receptors involved in control of prolactin secretion. Eur J Pharmacol 76:9–13, 1981

Rafaelsen OM, Clemmesen L, Lund H, et al: Comparison of peripheral anticholinergic effects of antidepressants: dry mouth. Acta Psychiatr Scand 63 (suppl 290):364–369, 1981

Rainbow TC, Biegon A: Quantitative autoradiography of [$^3$H]nitro-imipramine binding sites in rat brain. Brain Res 262:319–322, 1983

Reebye PN, Yiptong C, Sa-oon J, et al: A controlled double-blind study of femoxetine and amitriptyline in patients with endogenous depression. Pharmacopsychiatry 15:164–169, 1982

Richelson E: Tricyclic antidepressanö and neurotransmitter receptors. Psychiatr Ann 9:186–195, 1979

Ritvo ER, Freeman BJ, Yuwiler A, et al: Fenfluramine treatment of autism: UCLA collaborative study of 81 patienö. Psychopharmacol Bull 22:133–140, 1986

Ross SB, Åberg-Wistedt A: Inhibitors of serotonin and noradrenaline uptake in human plasma after withdrawal of zimelidine and clomipramine treatment. Psychopharmacology (Berlin) 79:298–303, 1983

Rowland NE, Carlton J: Neurobiology of an anorectic drug: fenfluramine. Prog Neurobiol 27:13–62, 1986

Rudorfer MV, Scheinin M, Karoum F, et al: Reduction of norepinephrine turnover by serotonergic drug in man. Biol Psychiatry 19:179–193, 1984

Settle EC, Settle GP: A case of mania associated with fluoxetine. Am J Psychiatry 141:280–281, 1984

Shoulson I, Chase TN: Fenfluramine in man: hypophagia associated with diminished serotonin turnover. Clin Pharmacol Ther 17:616–621, 1975

Somni RW, Crismon ML, Bowden CL: Fluoxetine: a serotonin-specific, second-generation antidepressant. Pharmacology 7:1–15, 1987

Stark P, Fuller RW, Wong DT: The pharmacologic profile of fluoxetine. J Clin Psychiatry 46(No 3, suppl 2):7–13, 1985

Steiner W, Fontaine R: Toxic reaction following the combined administration of fluoxetine and L-tryptophan: five case reporö. Biol Psychiatry 2:1067–1071, 1986

Thomas DR, Nelson DR, Johnson AM: Biochemical effecö of the antidepressant paroxetine, a specific 5-hydroxytryptamine uptake inhibitor. Psychopharmacology (Berlin) 93:193–200, 1987

Träskman L, Åsberg M, Bertilsson L, et al: Plasma levels of chlorimipramine and its desmethylmetabolite during treatment of depression. Clin Pharmacol Ther 26:600–610, 1979

Trulson ME, Jacobs BL: Behavioral evidence for the rapid release of CNS serotonin by PCA and fenfluramine. Eur J Pharmacol 36:149–154, 1976

Turner SM, Jacob RG, Beidel DC, et al: Fluoxetine treatment of obsessive-compulsive disorder. J Clin Psychopharmacol 5:207–212, 1985

van Praag HM, Kahn R, Asnis GM, et al: Therapeutic indications for serotonin-potentiating compounds: a hypothesis. Biol Psychiatry 22:205–212, 1987

Wernicke JF: The side effect profile and safety of fluoxetine. J Clin Psychiatry 46(No 3, suppl 2):59–67, 1985

Wong DT, Bymaster FP, Reid LR, et al: Fluoxetine and two other serotonin uptake inhibitors without affinity for neuronal receptors. Biochem Pharmacol 32:1287–1293, 1983

# Chapter 11

## *Clinical Significance of Central Serotonergic System Dysfunction in Major Psychiatric Disorders*

Emil F. Coccaro, M.D.
Dennis L. Murphy, M.D.

# Chapter 11

## *Clinical Significance of Central Serotonergic System Dysfunction in Major Psychiatric Disorders*

Consideration of the data reviewed in this volume suggests that while the behavioral syndromes examined herein are quite different in many respects, central serotonergic system dysfunction may be present in many patients with any of a variety of major psychiatric disturbances.

Evidence for decreased or increased central serotonin (5-hydroxytryptamine [5-HT]) function is present for many of these disorders. Evidence for decreased central 5-HT function is present for patients with autism, mood disorder, certain personality disorders, eating disorder, and alcoholism, while evidence for increased central 5-HT function is present for patients with obsessive-compulsive disorder (OCD) and perhaps other anxiety disorders as well.

While central 5-HT function appears to be abnormal in a wide variety of patients, evidence suggesting a more dimensional approach to the role of central 5-HT in psychopathology is becoming increasingly available. Inverse relationships between lumbar CSF 5-hydroxyindoleacetic acid (5-HIAA) concentrations, as well as neuroendocrine responses to central serotonergic probes, and behaviors such as suicide, self-damaging acts, impulsive aggression, and increased motoric activity, have been repeatedly demonstrated in patients with mood disorder, personality disorder, and schizophrenia as well as in healthy volunteers. Associations between increased central serotonergic activity and obsessive-compulsive anxiety and positive relationships between CSF 5-HIAA concentrations and deficit-state schizophrenic symptoms are now being reported. Similar data from CSF 5-HIAA studies in the nonhuman primate further indicate that developmental changes in central 5-HT function are correlated with developmental changes in behavior that appear to lie along a behavioral dimension of disinhibition (e.g., "irritable" impulsive aggression) at the low end and behavioral inhibition (e.g., anxiety, timidity) at the high end.

Overall, these data are consistent with similar data in more primitive species that suggest that the central serotonergic system is involved in a complex behavioral system mediating "harm avoidance" in which central 5-HT inhibits the organism from behaviors that have a high potential for punishment or adverse consequences. In this model, low central serotonergic activity is associated with a release of inhibition, so that the organism behaves in a manner that is repeatedly associated with an adverse outcome. Conversely, high central serotonergic activity is associated with an overinhibition of behavior, so that the organism avoids much activity, even that associated with a low potential of an adverse outcome. Under this model, we can begin to understand how low central 5-HT function may be associated with the suicidal acts of some patients with severe depression, schizophrenia, or alcoholism; the self-damaging acts of patients with autism or borderline personality disorder; and the impulsive violent acts of patients with borderline and/or antisocial personality disorder. Further, we get a clue as to how high central 5-HT function could be associated with the severe anxiety observed in patients with OCD.

While heuristic, this model is clearly simplistic and not capable of explaining all the clinical phenomena that may occur in patients with abnormalities of central 5-HT function. The model does not explain how schizophrenic patients with low central 5-HT function have increased motor activity or a greater number of first-rank Schneiderian symptoms while in an exacerbated stage of illness. Nor does it explain the occurrence of suicidal acts in patients with bipolar illness, with whom indices of central 5-HT function appear to be unassociated with history of suicidal acts.

Clearly then, there must be neurobiological differences among patients with abnormalities of central 5-HT function that account for the phenomenological differences observed clinically. Patients with depression and low central serotonergic activity may not have a history of suicidal or irritable aggressive acts, or of bulimia. Patients with eating disorders and low central serotonergic activity may not have concurrent depression or a history of suicidal or irritable aggressive acts. Similarly, patients with alcoholism and low central serotonergic activity may not have concurrent bulimia, depression, or a history of suicidal or impulsive aggressive acts.

These differences may possibly be due to selective "lesioning" within the central serotonergic system, either in individual patients or in patients with specific behavioral disorders. Since it is highly probable that the psychobiologic assessment strategies available only reflect limited areas of function subserved by central 5-HT, it is possible that differences within the central serotonergic system among patients with different disorders cannot be identified at this time. Clearly there are differences between patients with uncomplicated depression, borderline personality disorder, and autism, yet patients from all three groups demonstrate diminished prolactin responses to fenfluramine and appear indistinguishable on this assessment of 5-HT function in the hypothalamo-pituitary axis. However, it is possible that beneath this apparently similar dysfunction of (at least one aspect of) central serotonergic activity lie abnormalities of different 5-HT receptor subtypes, all of which may mediate prolactin release, but differentially mediate other, and perhaps critically different, neurobiological functions. Alternatively, it is possible that all these patients share the same abnormality of central serotonergic activity in the hypothalamo-pituitary axis but that there are differing abnormalities of other serotonergic neurons or of projections to differing areas of the brain that have not been, or cannot be, tested at this time. An example of this possibility is suggested by the heightened anxiety (i.e., presumably mediated by cortical structures), yet normative neuroendocrine (i.e., mediated by hypothalamo-pituitary structures), responses to *m*-chlorophenylpiperazine as observed in patients with OCD. Finally, differences may lie within other neurotransmitter (e.g., catecholaminergic, etc.) systems not reflected by the assessment under investigation. It is possible, for example, that many depressed patients with low central 5-HT function may not display prominent suicidal and/or irritable aggressive acts due to concomitant dysfunction of another system (e.g., norepinephrine, dopamine) necessary in order to mediate an impulse that might otherwise lead to a suicidal or aggressive act.

Potential differences among the neurobiological substrates of these disorders are more than of academic importance, however. While data from a variety of studies give insight into the psychobiologic underpinnings of the clinical response to empirically efficacious agents, our understanding of how to best treat many patients is incomplete.

Accordingly, future research must focus on at least two fronts. First, neuropharmacologic agents need to be developed that have primary, if not specific, actions on the central serotonergic system, as well as agents that have actions mediated by specific 5-HT receptor subtypes or actions localized to specific brain regions. Availability of such agents will be necessary in order to understand more fully the specific nature of the abnormalities in central 5-HT already detected by our crude assessments of this system. Results of such investigations may be used to guide the development of new approaches to mollify specific abnormalities of central 5-HT as may occur differently in these patients.

Second, there needs to be investigation into the roles played by other neurotransmitter systems that interact with central 5-HT and potentially alter the substrate upon which central 5-HT acts. Assessment of such interactions may greatly impact upon the development of new treatments for these disorders. Thus if dysfunction of the central serotonergic system is linked, or is secondary, to dysfunction of another system, interventions that affect the latter system may be needed to produce the desired clinical response. Evidence that simultaneous dysfunction of several neurotransmitter systems may underlie poor responses to clinical treatment is suggested by data which indicate that nonresponders to antidepressant treatment do not demonstrate the usually observed relationships among 5-HT and catecholamine metabolites in CSF and that receptor changes in noradrenergic systems ordinarily observed with antidepressant treatment do not occur after the disruption of central serotonergic inputs.

While the past two decades of research into the central serotonergic system have produced many interesting and important clues into the psychobiology of many of the major psychiatric disorders, much work remains to be done. The data reviewed in this volume set the stage for future efforts that seek to develop a greater understanding of the biological factors involved, and for a more specific approach to the treatment of these disorders.

# Appendix

## Glossary Relevant to Serotonin

### Synthesis and Catabolism

Tryptophan: Precursor of 5-HT
Tryptophan hydroxylase: Converts tryptophan into 5-HTP
5-Hydroxytryptophan (5-HTP): Immediate precursor of 5-HT
Aromatic amino acid decarboxylase: Converts 5-HTP into 5-HT
5-Hydroxytryptamine (5-HT): Serotonin
Monoamine oxidase (MAO): Converts 5-HT into 5-HIAA
5-Hydroxyindoleacetic acid (5-HIAA): Major 5-HT metabolite

### Serotonin Receptor Subtypes

$5\text{-HT}_{1A\text{-}D}$ receptors: 5-HT receptors with high affinity for 5-HT
$5\text{-HT}_2$ receptors: 5-HT receptors with high affinity for spiperone
$5\text{-HT}_{3A\text{-}C}$ receptors: 5-HT receptors with high affinity for 5-HT but not for the typical 5-HT antagonists (e.g., metergoline, ketanserin, etc.)

### Serotonin Agonist Probe Candidates

Intravenous tryptophan: 5-HT precursor agent
5-HTP: 5-HT precursor agent
Fenfluramine: 5-HT releaser/uptake inhibitor
Clomipramine: 5-HT uptake inhibitor
*m*-Chlorophenylpiperazine (m-CPP): Moderate affinity for all 5-HT receptor subtypes
Buspirone (BuSpar): Partial agonist; high affinity for $5\text{-HT}_{1A}$ receptor subtype
Gepirone: Congener of buspirone; but nearly a full 5-HT agonist
Ipsapirone: Congener of buspirone; but nearly a full 5-HT agonist
MK-212: Moderate affinity for 5-HT receptors in general; some responses in rodents (e.g., ACTH, temperature) appear to reflect $5\text{-HT}_2$ involvement

**Serotonin Antagonist Probe Candidates**

Metergoline: Nonselective 5-HT postsynaptic receptor antagonist
Methysergide: Nonselective 5-HT postsynaptic receptor antagonist
Pindolol: Nonselective beta-blocker with high affinity for the 5-$HT_1$ receptor subtype
Ketanserin: Relatively selective for the 5-$HT_2$ receptor subtype, with some alpha blockade
Ritanserin: Relatively selective for the 5-$HT_2$ and 5-$HT_{1C}$ receptor subtypes
MDL 7222/ICS 205 930: Selective for the 5-$HT_3$ receptor subtype

**Serotonin Agents Available for Therapeutics**

Trazodone (Desyrel): Antidepressant; precursor of m-CPP
Fluoxetine (Prozac): Antidepressant; selective 5-HT uptake inhibitor
Buspirone (BuSpar): Anxiolytic; high affinity for the 5-$HT_{1A}$ receptor subtype
Clomipramine (Anafranil): Antiobsessional/antidepressant; relatively selective 5-HT uptake inhibitor

**Serotonergic Agents Under Development for Therapeutics**

Fluvoxamine: Selective 5-HT uptake inhibitor
Sertraline: Selective 5-HT uptake inhibitor
Gepirone: Selective 5-HT agonist; congener of buspirone
Ipsapirone: Selective 5-$HT_{1A}$ agonist; congener of buspirone
Eltoprazine: Selective 5-$HT_{1A}$ and 5-$HT_{1B}$ agonist
Risperidone: Selective 5-$HT_2$ and $D_2$ antagonist